MYP Physical and Earth Sciences

A concept-based approach

Years 1–3

William Heathcote

OXFORD UNIVERSITY PRESS

OXFORD
UNIVERSITY PRESS

Great Clarendon Street, Oxford, OX2 6DP, United Kingdom

Oxford University Press is a department of the University of Oxford. It furthers the University's objective of excellence in research, scholarship, and education by publishing worldwide. Oxford is a registered trade mark of Oxford University Press in the UK and in certain other countries

First published in 2019

British Library Cataloguing in Publication Data
Data available

978-0-19-836998-1

10 9 8 7 6 5 4 3 2 1

Paper used in the production of this book is a natural, recyclable product made from wood grown in sustainable forests.
The manufacturing process conforms to the environmental regulations of the country of origin.

Printed in China by Golden Cup Printing Co Ltd

Acknowledgements
The author would like to thank the editors, Ben Rout and Jane Glendening, for all of their invaluable contributions and suggestions without which this book would not have been possible.

The publisher and authors would like to thank the following for permission to use photographs and other copyright material:

Cover: robertharding/Alamy Stock Photo. All other photos © Shuttertock, except **p3(b)**: Stephen Durke/OUP; **p4(t)**: © Getty Images; **p9**: Kevin Wheal/Alamy Stock Photo; **p11(t)**: Gallo Images/ Getty Images; **p13(t)**: Eduardo Ripoll/StockimoNews/Alamy Stock Photo; **p13(mb)**: Action Plus Sports Images/Alamy Stock Photo; **p13(b)**: dpa picture alliance/Alamy Stock Photo; **p19**: OUP; **p22(t)**: National Geographic Image Collection/Alamy Stock Photo; **p24(t)**: Gabriel Petrescu/Alamy Stock Photo; **p24(b)**: Stephen Barnes/ Industry and Engineering/Alamy Stock Photo; **p27(l)**: Edward Fullick and Tech-Set Ltd/OUP; **p34**: mauritius images GmbH/Alamy Stock Photo; **p36**: MediaWorldImages/Alamy Stock Photo; **p40(l)**: Stuart Boulton/Alamy Stock Photo; **p41**: Keystone Press/Alamy Stock Photo; **p42(t)**: Kevin Foy/Alamy Stock Photo; **p43**: Courtesy of the author; **p52(t)**: Science History Images/Alamy Stock Photo; **p52(t)**: US Government/Wikimedia Commons; **p54(l)**: The Natural History Museum/Alamy Stock Photo; **p56(tl)**: Phil Degginger/Alamy Stock Photo; **p56(tr)**: Photolibrary; Nigel Francis/Robert Harding Travel; **p61(b)**: Courtesy of the author; **p63(b)**: Science History Images/ Alamy Stock Photo; **p68(t)**: Xinhua/Alamy Stock Photo; **p69(t)**: Courtesy of the author; **p70**: PJF Military Collection/Alamy Stock Photo; **p72(b)**: Ashley Cooper/Alamy Stock Photo; **p79**: Courtesy of the author; **p89(b)**: John Gooday/Alamy Stock Photo; **p91(t)**: All Canada Photos/Alamy Stock Photo; **p93(b)**: Courtesy of the author; **p94, 96**: NASA Image Collection/Alamy Stock Photo; **p97(t, ml, mr)**: Courtesy of the author; **p103(t)**: Courtesy of the author; **p107(b), 112(b), 113(t)**: © Getty Images; **p115**: momentimages/Tetra Images/ Corbis; **p116**: ORNL/SCIENCE PHOTO LIBRARY; **p125(r)**: © Getty Images; **p134(t)**: Public Domain/Wikimedia Commons; **p135(t)**: © Corbis; **p136(l)**: © Getty Images; **p137(r)**: The Natural History Museum/Alamy Stock Photo; **p140(t)**: Oxford University Press ANZ/ Brent Parker Jones; **p141(tl)**: © Getty Images; **p141(br)**: SCIENCE PHOTO LIBRARY; **p145**: GIPHOTOSTOCK/SCIENCE PHOTO LIBRARY; **p148(t)**: Courtesy of the author; **p156(b)**: INTERFOTO/Alamy Stock Photo; **p157(t)**: Design Pics Inc/Alamy Stock Photo; **p158(t)**: World History Archive/Alamy Stock Photo; **p160**: SCIENCE PHOTO LIBRARY; **p163(b)**: SCIENCE PHOTO LIBRARY; **p170(b), 171(b)**: ANDREW LAMBERT PHOTOGRAPHY/SCIENCE PHOTO LIBRARY; **p172**: DORLING KINDERSLEY/UIG/SCIENCE PHOTO LIBRARY; **p174**: Pawel Burgiel/Alamy Stock Photo; **p181(t)**: © Getty Images; **p186**: JOEL AREM/SCIENCE PHOTO LIBRARY; **p190**: © Getty Images; **p194(b)**: Science History Images/Alamy Stock Photo; **p195(t)**: Science Photo Library/Alamy Stock Photo; **p195(b)**: RGB Ventures/ SuperStock/Alamy Stock Photo; **p200(b)**: Keystone Press/Alamy Stock Photo; **p201(t)**: Science Photo Library/Alamy Stock Photo; **p202(t)**: The Natural History Museum/Alamy Stock Photo; **p206(t)**: SPL Creative/Getty Images; **p206(b)**: © Corbis; **p224(l)**: World History Archive/Alamy Stock Photo; **p224(r)**: J Marshall - Tribaleye Images/ Alamy Stock Photo; **p225**: NASA; **p226**: dpa picture alliance archive/ Alamy Stock Photo; **p227**: NASA/Wikimedia Commons; **p228(l)**: European Southern Observatory (ESO)/Wikimedia Commons/CC BY 4.0; **p229**: NASA/Wikimedia Commons.

Artwork by QBS Media Services Inc.

Contents

Introduction

The MYP Physical and Earth Sciences course, like all MYP Sciences, is inquiry based. To promote conceptual understanding, the MYP uses key concepts and related concepts. Key concepts represent big ideas that are relevant across disciplines. The key concepts used in MYP Sciences are change, relationships and systems. Related concepts are more specific to each subject and help to promote more detailed exploration. Each chapter is focused on a topic area in Physical and Earth Sciences, one key concept and two of the 12 related concepts.

The 12 chapters in this book do not form a fixed linear sequence. They form a 3×4 matrix, organized by key concept:

Change	Systems	Relationships
Motion	Balanced forces	Unbalanced forces
Potential energy, kinetic energy and gravity	Heat and light	Waves and sound
Chemical reactions	Atoms, elements and compounds	Properties of matter
The Earth	The atmosphere	The universe

There are many different ways of navigating through this matrix. The ideal route will depend on students' ages and any additional requirements from the local science curriculum.

The objectives of MYP Science are categorized into four criteria, which contain descriptions of specific targets that are accomplished as a result of studying MYP Science:

A. Knowing and understanding

B. Inquiring and designing

C. Processing and evaluating

D. Reflecting on the impacts of science

Within each chapter, we have included activities designed to promote achievement of these objectives, such as experiments and data-based questions. We have also included activities designed to promote development of approaches to learning skills. The summative assessment found at the end of each chapter is framed by a statement of inquiry relating the concepts addressed to one of the six global contexts and features both multiple-choice questions and questions that require longer answers.

Overall, this book is meant to guide a student's exploration of Physical and Earth Sciences and aid development of specific skills that are essential for academic success and getting the most out of this educational experience.

How to use this book

To help you get the most out of your book, here is an overview of its features.

Concepts, global context and statement of inquiry

The key and related concepts, the global context and the statement of inquiry used in each chapter are clearly listed on the introduction page.

Activities

A range of activities that encourage you to think further about the topics you studied, research these topics and build connections between physical and earth sciences and other disciplines.

Vocabulary features are designed to introduce and familiarize you with the key terms you will need to know when studying the physical and earth sciences.

Experiments

Practical activities that help you develop skills for assessment criteria B & C.

Data-based questions

These questions allow you to test your factual understanding of physical and earth sciences, as well as study and analyse data. Data-based questions help you prepare for assessment criteria A, B & C.

ATL Skills

These approaches to learning sections introduce new skills or give you the opportunity to reflect on skills you might already have. They are mapped to the MYP skills clusters and are aimed at supporting you become an independent learner.

Summative assessment

There is a summative assessment at the end of each chapter; this covers all four MYP assessment criteria.

Mapping grid

This table shows you which key concept, related concepts, global context and statement of inquiry guide the learning in each chapter.

Chapter	Key concept	Related concepts	Global context	Statement of inquiry	ATL skills
1 Motion	Change	Movement Patterns	Orientation in space and time	Knowing our position in space and time helps us to understand our place in the world.	**Communication skills**: Using mathematical notation **Research skills:** Understanding intellectual property
2 Balanced forces	Systems	Interaction Function	Identities and relationships	The interaction of forces can create a balanced system.	**Communication skills**: Communicating numerical quantities
3 Unbalanced forces	Relationships	Energy Consequences	Scientific and technical innovation	The relationship between unbalanced forces and energy has enabled huge improvements in technology.	**Transfer skills**: Isaac Newton
4 Potential energy, kinetic energy and gravity	Change	Form Energy	Orientation in space and time	Changes in energy drive the basic processes of nature.	**Critical thinking skills**: Observing carefully in order to recognize problems
5 Waves and sound	Relationships	Models Interaction	Personal and cultural expression	The waves that we see and hear help to form our relationship with the outside world.	**Communication skills**: Understanding human perception
6 Heat and light	Systems	Environment Development	Globalization and sustainability	Our environment is governed by the behavior of heat and light.	**Self-management skills:** Emotional management

Chapter	Key concept	Related concepts	Global context	Statement of inquiry	ATL skills
7 Atoms, elements and compounds	Systems	Patterns Models	Identities and relationships	The complex chemicals that enable life to exist are formed from only a few different types of atom.	**Communication skills**: Organizing and depicting information logically
8 Properties of matter	Relationships	Form Transformation	Globalization and sustainability	New materials with differing properties have helped to create today's global society and may hold the answers to some of the problems of the future.	**Communication skills**: Technology and collaboration
9 Chemical reactions	Change	Function Evidence	Fairness and development	As people have used chemical reactions, their society has changed and developed.	**Communication skills**: Choosing a line of best fit
10 The Earth	Change	Transformation Development	Fairness and development	The Earth has been transformed by slow processes that still continue today.	**Creativity and innovation skills**: Earthquakes in cities
11 The atmosphere	Systems	Consequences Environment	Globalization and sustainability	The atmosphere around us creates the conditions necessary for life.	**Communication skills**: International cooperation
12 The universe	Relationships	Evidence Movement	Orientation in time and space	The study of our solar system and the wider universe can lead to a better understanding of our own planet.	**Communication skills**: Making contact

1 Motion

Key concept: Change

Related concepts: Movement, Patterns

Global context: Orientation in space and time

◀ A marathon is a race over 42.195 km. Because of the large distance, marathons are tough, but they can be popular and large marathons can attract tens of thousands of runners. Elite runners can complete the course in about two hours, other runners may take 3 or 4 times as long. Why do tough challenges attract competitors?

▼ The global positioning system (GPS) uses satellites to provide information that allows a GPS device to measure its position to an accuracy of less than a meter. The technology has allowed the development of navigation systems and could allow driverless cars to function in the future. As well as helping emergency services, airlines, shipping and the military with navigation, the satellite system has been used to improve weather forecasting and synchronize computer timing for economic transactions. How else is modern life dependent on technology?

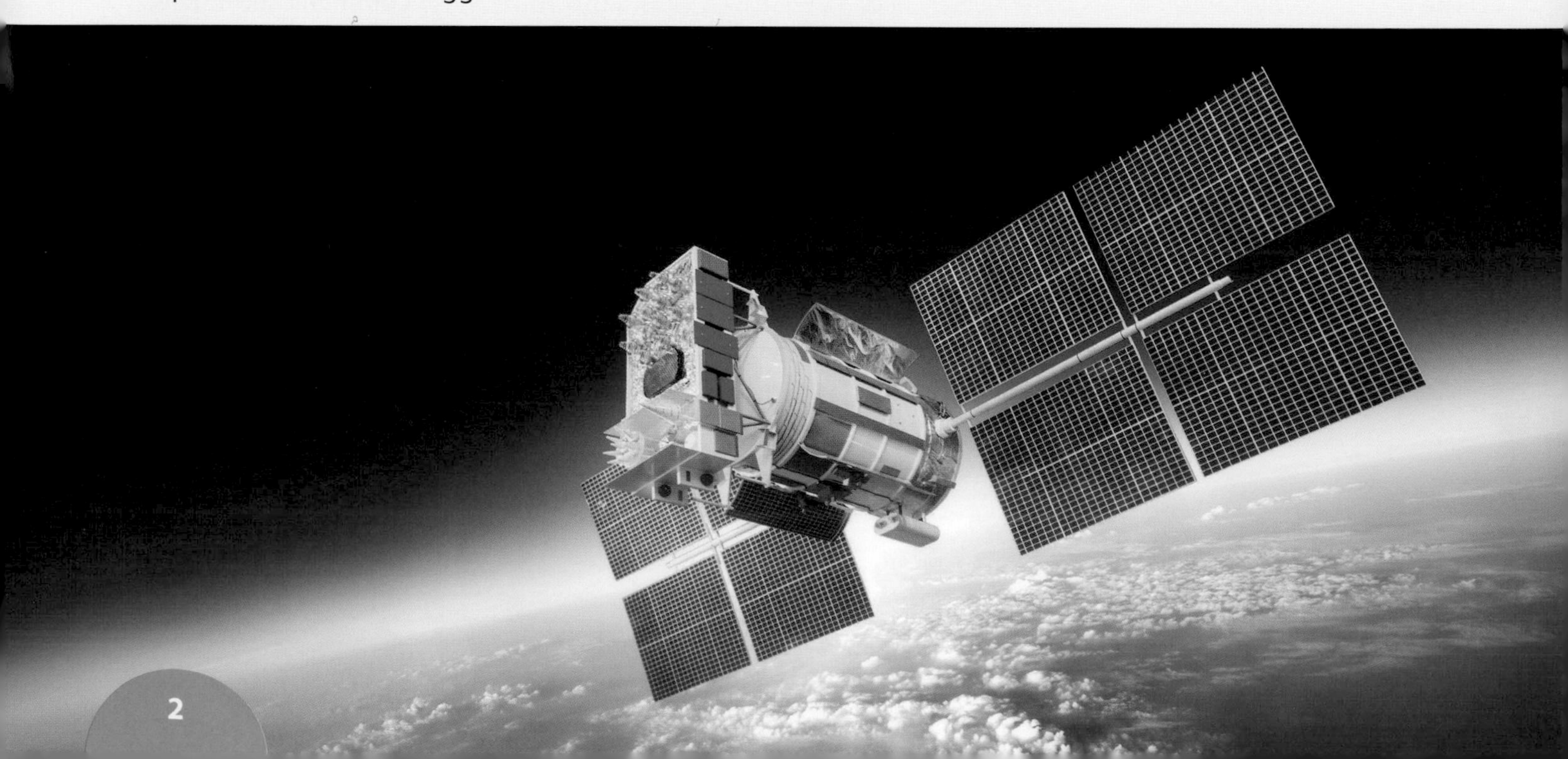

▲ It is not just animals that are capable of moving: plants can move in a variety of ways. The Venus flytrap (*Dionaea muscipula*) can close its leaves in under a second, allowing it to trap insects. How else do plants move?

◀ The once-straight strata in these rocks have been folded over millions of years by the motion of tectonic plates (See Chapter 10, The Earth). The motion of these plates—also called continental drift—happens so slowly (a couple of centimeters per year) that the theory of continental drift was not widely accepted until the late 1950s. What other things move in a way that is imperceptible?

Key concept: Change

Related concepts: Movement, Patterns

Global context: Orientation in space and time

Statement of inquiry:

Knowing our position in space and time helps us to understand our place in the world.

Introduction

Everything moves in some way. Whether it is the tiny atoms in a solid that are vibrating (see Chapter 8, Properties of matter) or the motion of the Earth around the Sun, every object moves or has parts of it that move.

As things move, they change. They may change location, speed or even their nature. Knowledge of how the things around us are moving and changing helps us to understand the world and its past, present and future. The key concept of this chapter is change and the global context of the chapter is orientation in space and time.

In this chapter we will see how we can measure motion and speed, and how we can use these measurements to predict an object's future motion. We will also see how acceleration can cause a change in motion.

Some things move quickly, others slowly. They may move backwards and forwards, in straight lines, or in circles. Knowing how an object moves allows us to know where an object was, or will be, and possibly predict when things will happen. We will use graphs to model motion, and so the related concepts are movement and patterns.

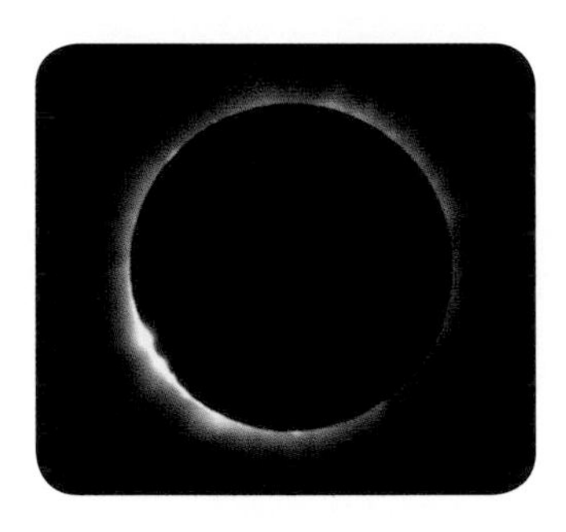

Whenever the positions of the Earth, Moon and Sun align, the moon blocks the Sun's light and the Earth passes through the shadow. As a result, there is a solar eclipse. Because the motion of the Moon and Earth are highly predictable, eclipses can be predicted thousands of years in advance. How predictable or unpredictable is the motion of other objects?

This picture shows an artist's impression of ʻOumuamua, an object that was discovered in 2017. It is hundreds of meters long, but much narrower. At first it was thought to be a comet or an asteroid, but analysis of its motion showed that it did not come from our solar system. Instead, it is the first known interstellar object—something that has come from somewhere else in our galaxy. Its closest approach to the Sun was closer than Mercury's orbit and since then it has been traveling away from the Sun at about 26 km/s

In 1851, Léon Foucault, a French physicist, installed a pendulum in the Panthéon in Paris. It was used to demonstrate the motion of the Earth. The pendulum swings back and forth, but as the Earth rotates, it causes the direction in which the pendulum oscillates to change. How else can the rotation of the Earth be shown?

How can we describe an object's position?

Movement involves the change in **position** of an object and so we must be able to describe its position if we are going to be able to describe its motion.

You will probably be able to see many things around you right now. This book is probably quite close to you and in front of you. Other things will be further away, maybe on either side of you or behind you. To describe the position of these objects, we need a fixed position to measure from (this can be your position) and a unit of length in which we will measure the **distance**. There are many units of distance that can be used. The SI unit of length (the scientific standard) is the **meter**. We will also need to consider direction. Normally, we would use positive numbers to describe the distance to objects in front of you and negative numbers to describe things behind you.

Quantities such as position (also called displacement), where the direction is important, are called **vector** quantities. Other quantities where a direction is not required, like distance, are called **scalar** quantities.

ABC

The **position**, or **displacement**, of an object is a measure of its distance from a fixed point.

Distance is a measurement of the length between two points. The SI unit of distance is the **meter**.

A **scalar** quantity is one that has magnitude (size), but not direction.

A **vector** quantity is one that has both magnitude and direction.

Units of distance

There are many different units of distance. Some of them are shown in the table below.

Unit	Symbol	Distance in meters
Millimeter	mm	0.001
Centimeter	cm	0.01
Kilometer	km	1000
Inch	"	0.0254
Mile	mi	1609.344
Light year	ly	9.46 thousand million million

1. Why do you think we use kilometers and miles to measure distances between towns?
2. Use the table to find how many inches are in a mile.
3. What sorts of distances might be measured in light years?
4. Calculate your height in miles.

ATL **Communication skills**

Using mathematical notation

The distances between galaxies are vast and are often measured in millions of light years. Other distances, such as the size of an atom, are tiny. To express these large or small distances in meters is not easy. Instead, scientists use standard form. This uses powers of ten where the power (the small number above the ten) gives the number of zeros. As a result, one thousand (1,000) is written as 10^3 and one million (1,000,000) is 10^6. A number such as 53,000 would be written as 5.3×10^4, you can think of the decimal point having moved 4 places from the end of 53,000 to between the 5 and the 3.

▲ The pinwheel galaxy is about two hundred thousand million million million meters away. This can be written as 2×10^{23} m

A negative power of ten means that the decimal point has moved the other way. Therefore a negative power gives a number less than 1. One thousandth is written as 10^{-3} and one millionth is 10^{-6}. A number such as 0.00275 would be written 2.75×10^{-3} because the decimal point has moved to the right by three places.

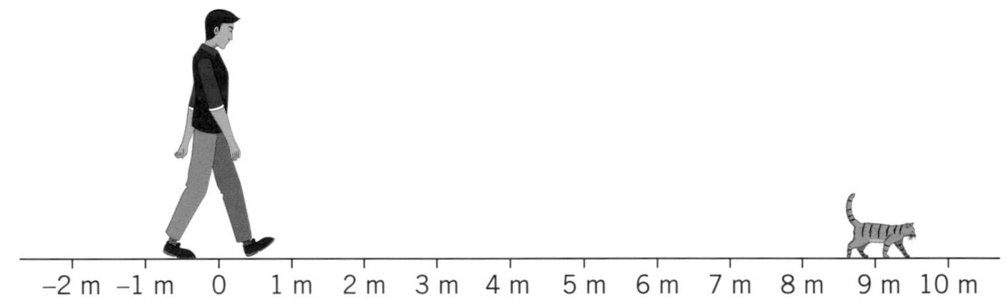

▲ Using a distance scale helps us to determine an object's position. Here, the cat is 9 meters away from the person. If the person is stationary and the cat gets further away or closer, then we would know that it is moving. Note that it is possible for the cat to be behind the person—in that case, we might use a negative number to show this

To account for objects that are not directly in front of you or behind you, we need a two-dimensional grid. In this case we use two coordinates that are often called x and y, but could also be North/South and East/West. The two axes are at right angles to one another and by measuring the position against each axis, an object's position can be recorded.

If we wish to find the distance between two objects in two dimensions, we need to use Pythagoras's theorem. Because the two axes are at right angles, the x and y coordinates form the two smaller sides of a right-angled triangle. The distance of the long side of the triangle can be found by using the equation:

$$\text{distance}^2 = x^2 + y^2$$

ATL **Research skills**

Understanding intellectual property

The ancient Greek mathematician and philosopher Pythagoras is credited with the discovery of the relationship between the lengths of the sides of a right-angled triangle, although it is likely that the relationship was known hundreds of years before by Babylonian, Mesopotamian and Indian mathematicians.

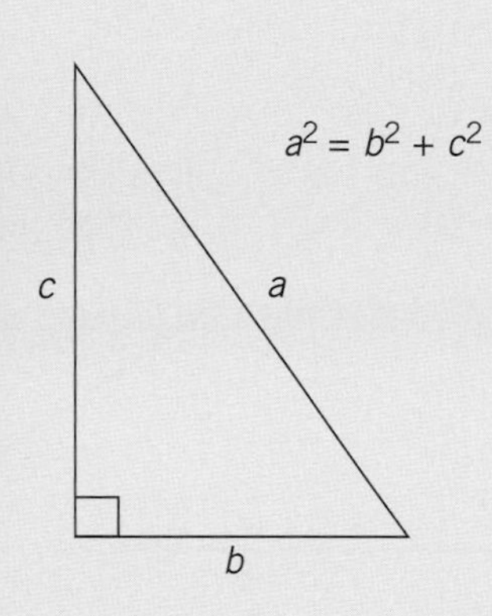

Trying to accurately associate discoveries with the correct discoverer is difficult when the discoveries were made thousands of years ago. For more modern discoveries and scientific theories, it is important to establish who has developed them and who should be credited with them.

Today, there are various legal systems that protect intellectual property, such as the patent system. This protects people and businesses who have developed a product by stopping other people from selling that product.

When publishing scientific papers, scientists often build on the ideas that have been developed by others. This is an important way in which scientific progress is made. Although scientists welcome other scientists using their ideas, they also want to be credited with their ideas. This is why scientific papers will reference papers by other scientists.

To reference a book, you should name the author, the title of the book, the year it was published and the publisher. It is sometimes useful to give the page number too. For example, a reference to this book would be: W Heathcote, **MYP Physical and Earth Sciences 1–3**, *Oxford University Press* (2019).

A website can be referenced in a similar way. Instead of the publisher, the website address is given as well as the date on which the information was last accessed. It might not be known who the author is or when the website was written, but if that information is available then it should be given. For example, the reference in the summative assessment at the end of this chapter is: World Health Organization, 2019, **WHO Global Ambient Air Quality Database (update 2018)**. *https://www.who.int/airpollution/data/cities/en/* (accessed 2 Apr 2019).

Normally, a scientific paper or book would have the references at the end in a section called the bibliography. At the point where the information appears in the paper, a reference would be given so that the reader knows where to look in the bibliography. This is normally given in square brackets and gives the author's surname and the year in which the source was written.
For example [Heathcote, 2019] or
[World Health Organization, 2019].

A compass is used in navigation. The compass needle always points north and this helps you to compare your surroundings to a map

▼ Monarch butterflies breed in the north east of America and Canada but every year they migrate thousands of kilometers to Mexico for the winter. After winter, they start migrating north again. As they travel north, they lay eggs that hatch into caterpillars. When these caterpillars pupate into butterflies, these new butterflies continue to travel north. It is not fully understood how the butterflies know their position and navigate

The two ideas of distance and position are slightly different to one another. Distance can be expressed with a single number. If you were to get a measuring tape and lay it out between two objects, you would be able to measure the distance between them. The position takes into account the direction in which an object is as well as its distance. As a result, you might describe a position as being 2.5 m *behind* you and 1 m to the right or with coordinates (–2.5, 1).

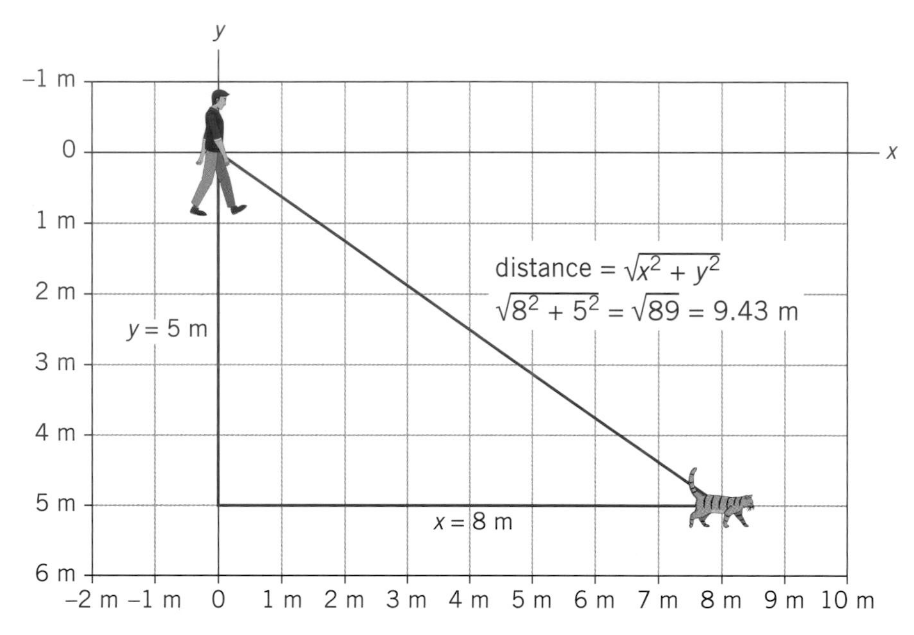

▲ Here, the position of the cat can be described as having an x coordinate of 8 m and a y coordinate of 5 m. This is often written as (8, 5). We can find the distance between the person and the cat by using Pythagoras's theorem, $\text{distance}^2 = x^2 + y^2$. In this case $\text{distance}^2 = 8^2 + 5^2 = 89$. Therefore the distance is $\sqrt{89}$, which is equal to 9.43 m

1. Two children are on a playing field. Their positions are measured in meters, relative to the gate of the field. The coordinates of the two children are (9, 40) and (33, 33).
   a) Draw a diagram showing their positions.
   b) How far away is the first child from the gate?
   c) What is the distance between the children?

2. A lighthouse keeper can see two boats at sea. One is 4 km due North and the other is 1 km North and 4 km to the East.
   a) Which boat is further from the lighthouse?
   b) Draw a diagram to show the position of the two boats.
   c) Calculate the distance between the two boats.

▲ Maps also use a grid system to display the position of buildings and landmarks

How do we describe motion?

An important concept in describing how something moves is its **speed**. Speed can be defined as the change in position divided by the time taken:

$$\text{speed} = \frac{\text{distance traveled}}{\text{time}}$$

Just as there was a difference between distance and position, where distance didn't consider the direction in which an object was, the quantity of speed does not account for the direction of motion. The quantity that is related to speed and includes the direction is called **velocity**.

To measure the speed and velocity in a single dimension, you must take account of how far the object has traveled and how long it takes to do this.

ABC

Speed is a measure of how much distance an object moves by in a given time. It is a scalar quantity.

Velocity is a vector quantity. It is also a measure of the distance moved by an object in a given time, but it has a direction as well.

Here, the cat moves from a distance of 5 meters to a distance of 9 meters in 8 seconds.

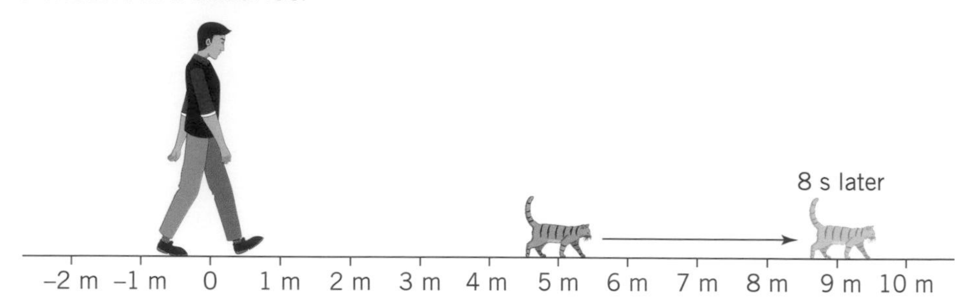

To calculate the speed, use the equation:

$$\text{speed} = \frac{\text{distance}}{\text{time}}$$

The distance traveled = 9 − 5 = 4 m, therefore

$$\text{speed} = \frac{4\text{ m}}{8\text{ s}} = 0.5\text{ m/s}$$

If the cat was traveling in the opposite direction, its speed would still be 0.5 m/s, but its velocity would be −0.5 m/s. The minus sign shows that the direction is backwards.

In two dimensions, the process of finding an object's speed is similar. Compare the starting position and the finishing position and find the difference. You may have to use Pythagoras's theorem to find the total distance traveled. You then divide by the time taken to find the speed.

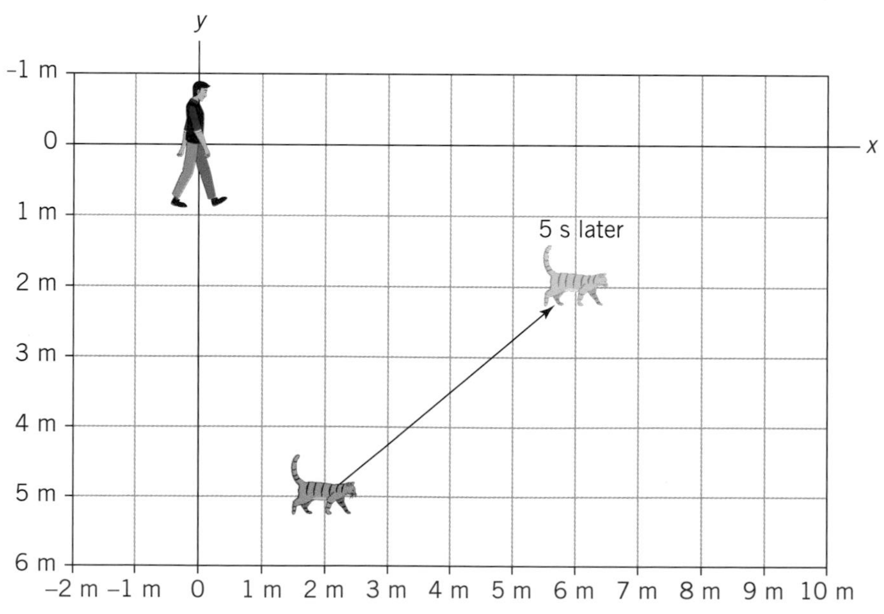

▶ In this picture, the cat starts at position (2, 5) and moves to position (6, 2). This is a change of 4 m in the x direction and 3 m in the y direction. We can use Pythagoras's theorem to find the total distance traveled: $\text{distance}^2 = 4^2 + 3^2 = 25$ and so the distance traveled is 5 m. Since this takes 5 s, the speed is 1 m/s

To find the cat's velocity we would need to consider the speed in both the x and y directions. In the x direction it moved 4 m. Using the equation above, this gives a speed of 0.8 m/s. In the y direction, it moved 3 m and so its speed is 0.6 m/s. Therefore, we can describe its velocity as (0.8, 0.6) m/s.

1. A car is 100 m from you traveling at 10 m/s. After 8 seconds:
   a) What is the furthest the car could be from you?
   b) What is the nearest the car could be to you?
   c) How might it be possible that the car is still 100 m away?
   d) In each case, how would you describe the car's speed and its velocity?

How can speed be measured?

When measuring speed, it is usual to measure a time and a distance so that the speed can be calculated. Often, a fixed distance is used, and the time taken to cover the distance is measured—this is essentially a race. It is possible that the speed might change throughout a race, so calculating the speed using the total distance and time will give the average speed for the race.

As with any experiment or measurement, it is important to control certain variables to make the race fair. Such control variables might be the distance over which the race is run; whether the racers have to be stationary at the start; or the method of movement or vehicle that is allowed.

Measuring speed in different races

Look at the following examples of races and answer the questions.

	1. The Olympic 100 m race is an example of a race over a fixed distance from a stationary start. This race was won in a time of 9.64 s. What was the average speed of the winner?
	2. The Le Mans 24 hours race is an example of a race where the time is fixed. The winner of this race completed 370 laps of the 13.6 km circuit. What was the average speed?
	3. To set a land speed record, the course is set as 1 mile long and a moving start is used. The car must complete two runs, one in each direction, within one hour. The recorded speed is the average of these two runs. Why does the course have to be run twice?

Experiment

Measuring speed

You can use a small battery powered car if you want to do this experiment inside, or you could run or cycle outside.

Method

- Mark a start line and a finish line, between which you will measure the speed. Measure the distance between the lines.
- Start the toy car a small distance before the start line and use a stopwatch to time from when it crosses the start line until it crosses the finish line. If you are going to run or cycle then you may need a friend to help you time.
- Repeat the experiment two more times.

Questions

1. Use the equation speed $= \frac{\text{distance}}{\text{time}}$ to calculate the speed.
2. If similar results are obtained for a measurement in an experiment, the results are said to be reliable. How reliable was your speed measurement? Consider whether you measured the same speed each time.
3. How could you improve your measurement of the speed such that your results are more reliable?

What are the units of speed?

Many units of speed are used. Fighter planes that fly faster than the speed of sound use Mach numbers, which is the speed of the plane relative to the speed of sound. Other planes and ships use units of speed called knots. More usually the units of speed involve a distance measurement and a time measurement, such as meters per second or miles per hour. In fact, a knot is a nautical mile per hour where a nautical mile is 1.15 miles or 1.85 km.

The speedometer on cars often shows the speed in miles per hour (mph) as well as kilometers per hour (km/h)

How far can they go?

The table shows the speed of some fast people, animals and vehicles. The examples given can maintain their speed for about an hour, but there are people, animals and cars that can travel faster than these over short distances.

Category	Notes	Speed	
Human	Abraham Kiptum broke the record for a half marathon (21.0975 km) with a time of 58 min 18 s in 2018.	6.03 m/s	
Animal	Although their speeds have not been measured for a full hour, the ostrich and the pronghorn antelope can both maintain a speed of about 30 mph for a long time.	30 mph (miles/hour)	
Cyclist	In 2019, Victor Campenaerts set the hour record by cycling 55.089 km in 1 hour.	55.089 km/h	
Car	Michael Schumacher won the 2003 Italian Grand Prix (306.72 km) in a time of 1 hour 14 minutes and 19.8 s.		

1. Calculate Michael Schumacher's average speed for the race in m/s. Remember to convert the time into seconds.
2. Convert the other speeds in the table into m/s. 1 mile is equal to 1609 m.
3. Calculate how far each example could travel in 1 second, 1 minute, 1 hour and 1 day. You may wish to give some of your answers in km.
4. Find a map that shows your location. Plot how far each example could travel in 1 hour and 1 day.

How can we find the average speed?

Average speed is a useful concept. It is defined as:

$$\text{average speed} = \frac{\text{total distance traveled}}{\text{total time taken}}$$

So, the average speed for a journey depends on the total distance traveled and the time taken rather than the speed at any given time.

How fast is Usain Bolt?

The table gives some of the split-times for Usain Bolt's 100 m world record.

Distance (m)	Time (s)
0	0
20	2.89
40	4.64
60	6.31
80	7.92
100	9.58

1. Plot a graph of the data with time on the x-axis and distance on the y-axis. Usually the first column of a table shows the independent variable, and this would be plotted on the x-axis. Here it is easier to see what is happening if time is plotted on the x-axis.
2. Which 20 m distance did Usain Bolt cover in the shortest time?
3. Calculate Usain Bolt's maximum speed based on your answer above.
4. What was Usain Bolt's average speed over the race?

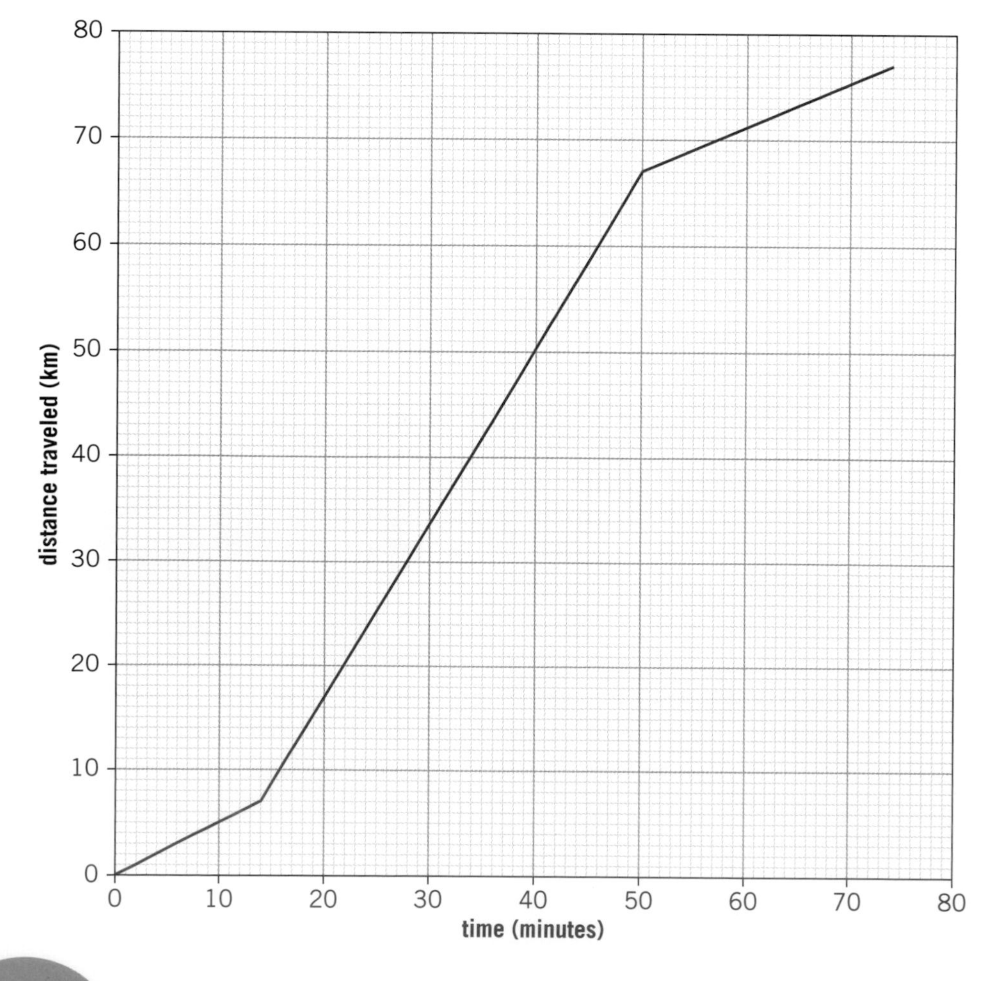

1. The graph shows a car journey between two towns. At first, when the car is in the town, it travels slowly. Once it is out of the town it speeds up. When it reaches the town it is traveling to, it slows down until it reaches its destination.
   a) Use the graph to find the average speed of the journey (give your answer in km/h).
   b) Find the car's speed when it is:
      i) in the first town
      ii) in the second town
      iii) between the towns.

What happens when an object changes speed?

Not all objects maintain the same speed or velocity. Some things get faster, others get slower. Since the concept of velocity includes direction, even a change in direction would mean that an object's motion has changed.

Whenever an object changes speed or velocity, we say that it accelerates. We define **acceleration** as being the change in speed or velocity divided by the time taken for this change to take place. This can be written as an equation:

$$\text{acceleration} = \frac{\text{change in velocity}}{\text{time taken}}$$

When a car brakes, it slows down and we might use the word **deceleration** to describe its motion. Because its speed is decreasing, the acceleration will be negative. If the car gets faster, its acceleration will be positive.

ABC

Acceleration is the rate at which an object's speed changes.

Deceleration is a special case of acceleration where the object is getting slower.

▼ Fighter pilots can experience very high accelerations (sometimes in excess of 50 m/s^2), particularly when the plane changes direction

1. Some salamanders have been found to be able to accelerate their tongues at an acceleration of 4,000 m/s^2. The acceleration only lasts for about 0.001 s. If their tongue is stationary to begin with, how fast is their tongue moving after this time?
2. The fastest acceleration in the animal kingdom is found in the mantis shrimp. It has specialized club-like limbs that it uses to strike its prey. It can accelerate these clubs to speeds of 23 m/s in a time of 0.0027 s. Calculate the acceleration of its clubs.

▲ A mantis shrimp can move its club-like limbs with a huge acceleration

How can we represent motion with a graph?

A simple way of representing motion is using a distance–time graph. These graphs normally plot the time on the *x*-axis and the position on the *y*-axis.

The following ideas will help you to interpret a distance–time graph:

- If the line is horizontal, it means that the position is not changing—the object must be stationary.
- If the line is sloping, the object must be moving. The steeper the slope, the faster it is moving.
- The gradient of the line represents the speed.
- If the line is not straight, then it means that the slope is changing—the object must be changing speed and so it is accelerating or decelerating.

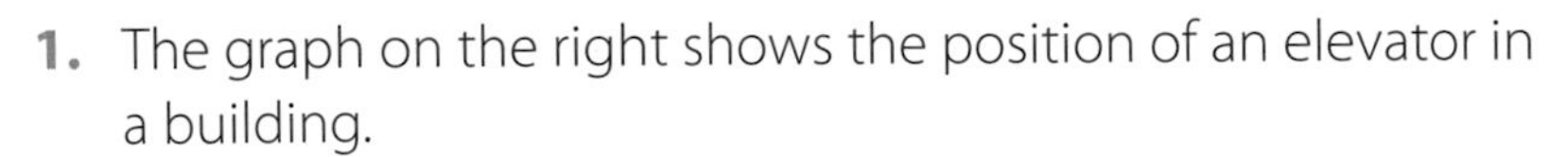

1. The graph on the right shows the position of an elevator in a building.
   a) When is the elevator stationary?
   b) When is the elevator moving fastest?
   c) Calculate the speed of the elevator over the first 4 seconds.
   d) What was the total distance traveled by the elevator?
   e) Explain why the total distance traveled is different to the final position of the elevator.

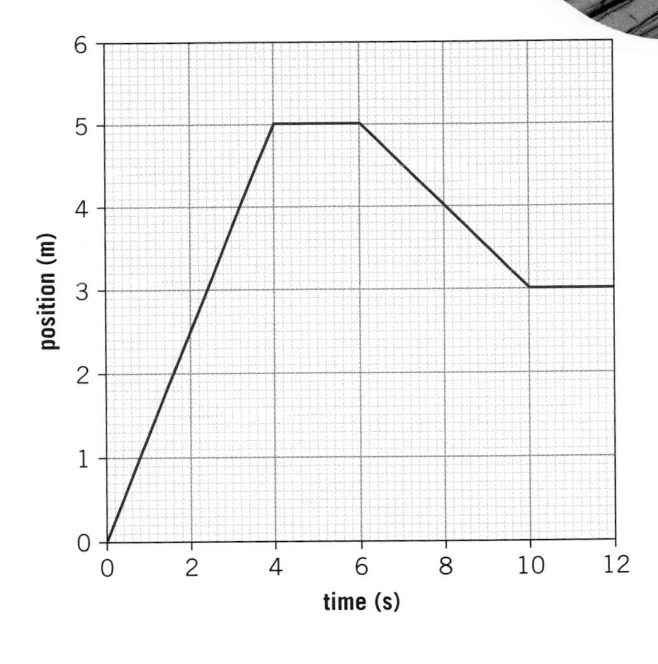

2. The graph shows the height of a toy rocket after it is launched.

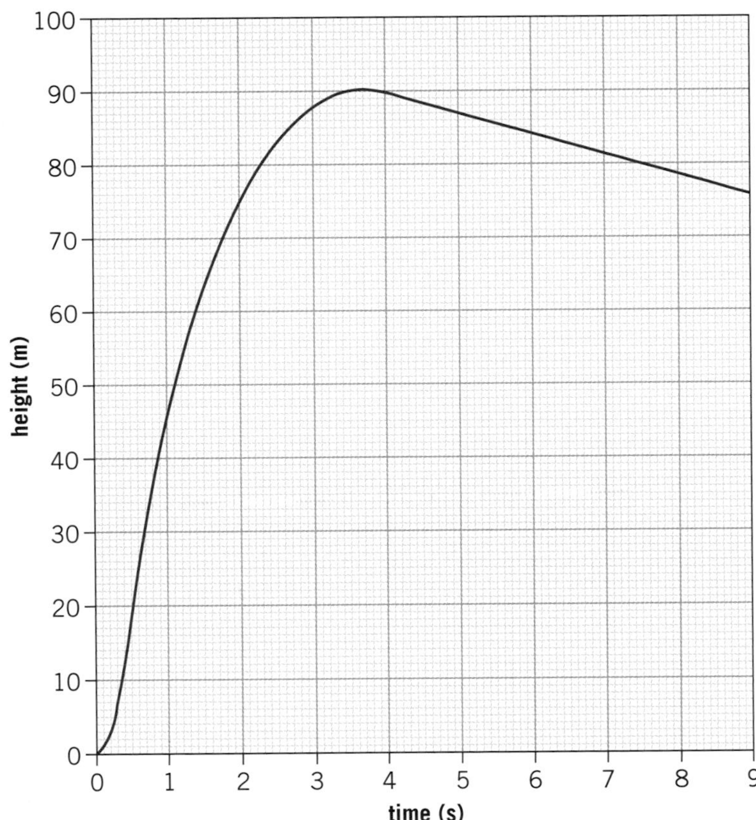

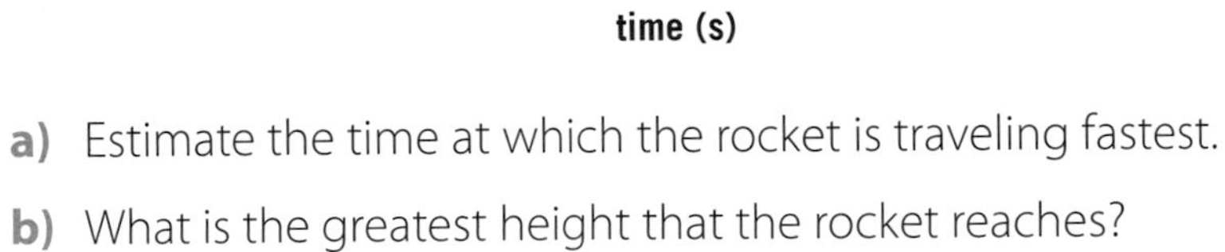

   a) Estimate the time at which the rocket is traveling fastest.
   b) What is the greatest height that the rocket reaches?
   c) Describe the motion of the rocket during the first 3.5 s.

   When the rocket reaches the top of its flight, it has a parachute that opens.

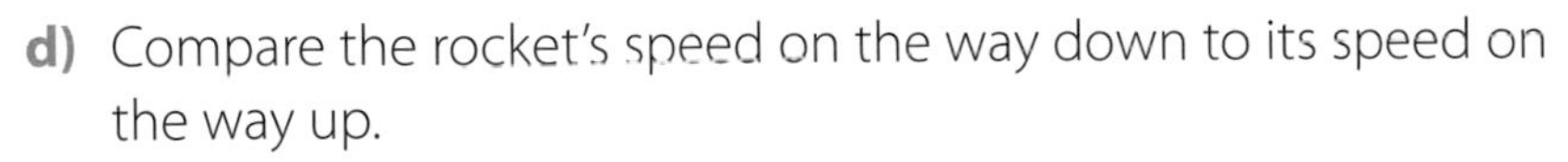

   d) Compare the rocket's speed on the way down to its speed on the way up.
   e) Use the graph to find the speed at which the rocket falls.
   f) Estimate the time after its launch that that the rocket will land.

Summative assessment

Statement of inquiry:

Knowing our position in space and time helps us to understand our place in the world.

This assessment is based around cycling as a sport as well as a mode of transport.

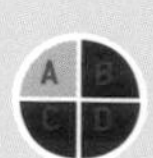

Cycling and motion

1. If a cyclist travels at an average speed of 15 km/h, how far can she go in 20 minutes?

   A. 0.5 km
   B. 5 km
   C. 30 km
   D. 300 km

2. A cyclist travels at 12 km/h for 5 km and then he turns around and cycles back 1 km. The whole journey takes 30 minutes. What was the cyclist's average speed?

   A. 0.5 km/h
   B. 2.5 km/h
   C. 8 km/h
   D. 12 km/h

3. A cyclist travels at a constant speed. This is the distance–time graph for her motion. What is the cyclist's speed?

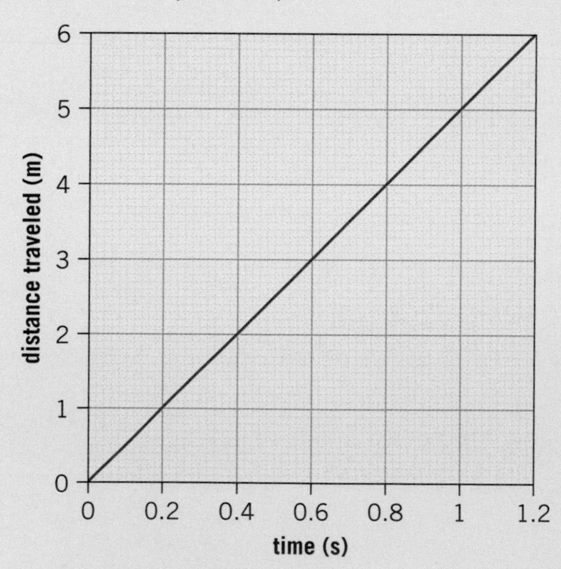

   A. 0.2 m/s
   B. 1.2 m/s
   C. 5 m/s
   D. 7.2 m/s

4. A cyclist has a puncture and is pushing his bike. Another person observes how far away the cyclist is. These results are plotted in the graph below. Which description best describes the cyclist's motion?

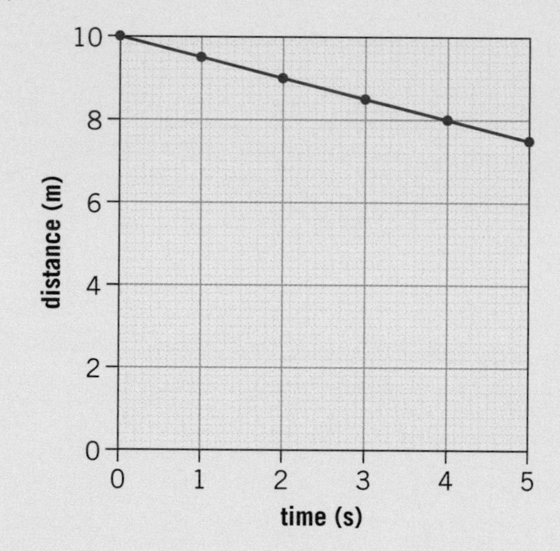

   A. Decelerating
   B. Accelerating
   C. Moving away at a constant speed
   D. Moving towards him with a constant speed

5. A cyclist rolls down a hill. He starts from rest and after 8 s he is traveling at 13.6 m/s. What is his acceleration?

   A. 1.7 m/s^2
   B. 5.6 m/s^2
   C. 21.6 m/s^2
   D. 108.8 m/s^2

6. If you cycle 1200 m North and then 1600 m West, how far are you from your starting point?

   A. 400 m
   B. 2,000 m
   C. 2,600 m
   D. 2,800 m

Choosing a bicycle

A cyclist wants to test four different bikes to see wthich is the fastest. To do this, he designs a simple course. He gets four friends to help him compare the bikes by racing them around the same course. The cyclist starts them at the same time and times how long it takes them to complete the course. His results are shown in the table below.

Bicycle	Time (minutes:seconds)
A	03:14
B	03:42
C	02:57
D	03:11

The cyclist then plots a graph of his data.

7. Which type of graph should the cyclist use to display his data? [1]
8. Identify the independent and dependent variables in the cyclist's experiment. [2]
9. Identify a control variable that the cyclist successfully controlled in his experiment. [1]
10. Suggest one variable that was not controlled in this experiment. [2]
11. Outline an improved method that the cyclist could use to decide which bike is fastest. [4]

▲ What factors other than speed might you consider when choosing a bike?

Cycling time trials

A cyclist wants to measure how fast he can cycle. He goes to a velodrome that has a 250 m track. He positions his bike on the start line at rest. He gets a friend to start a stopwatch when he starts pedaling and to measure the time that has elapsed each time he re-crosses the starting line.

The times are shown in the table.

Lap	Total time (minutes:seconds)
1	0:25
2	0:52
3	1:20
4	1:46
5	2:10

12. Calculate the cyclist's average speed over the first lap. [2]

13. Explain why it is likely that the cyclist crosses the line after the first lap at a higher speed than the one you calculated in the previous question. [2]

14. Explain why it can be said that the cyclist's average velocity over the first lap is zero. [3]

15. Use the table to find the cyclist's fastest lap and hence find the cyclist's fastest average speed over a lap. [2]

16. Explain why it is possible that the fastest speed that the cyclist attained was higher than your result. [1]

Health benefits of cycling and pollution

In 2016, a study by M. Tainio and others at King's College London compared the health benefits of cycling with the negative effects of exposure to air pollution in a city. A reference for this scientific paper can be given as the following:

M. Tainio et al., Can air pollution negate the health benefits of cycling and walking?, *Preventative Medicine*, **87**, pp. 233-236 (2016)

17. Give one reason why is it important to reference where this information has come from. [1]

The table gives the average pollution levels of some cities, shown by the concentration of $PM_{2.5}$ in these cities. $PM_{2.5}$ are very small particles, with a diameter of less than 2.5 micrometers. The unit of concentration is micrograms per cubic meter (μg/m^3). A microgram is a millionth of a gram. For example, in Johannesburg, there are 41 micrograms of $PM_{2.5}$ particles in every cubic meter of space.

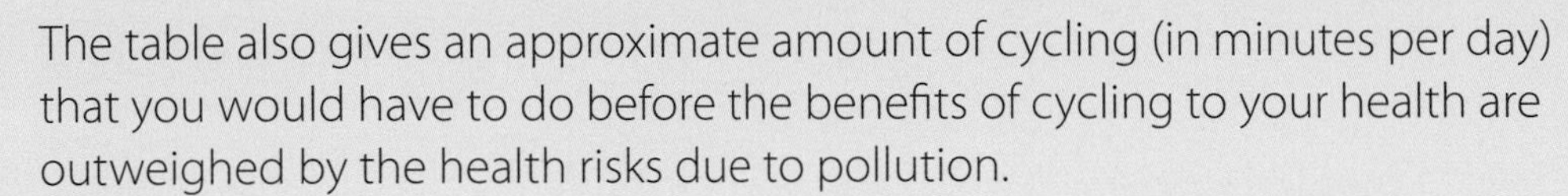

The table also gives an approximate amount of cycling (in minutes per day) that you would have to do before the benefits of cycling to your health are outweighed by the health risks due to pollution.

City	Pollution levels ($PM_{2.5}$ concentration in μg/m^3)	Amount of cycling before it becomes harmful (minutes per day)
Johannesburg	41	480
Cairo	117	60
Delhi	143	40
Beijing	73	150

Source of data: World Health Organization, 2019, **WHO Global Ambient Air Quality Database (update 2018)**. *https://www.who.int/airpollution/data/cities/en/* (accessed 2 Apr 2019)

18. The average cycling speed is 15 km/h. Calculate how far you would be able to cycle in each city before it becomes harmful. [4]

19. For pollution levels lower than 24 μg/m^3 the study suggests that the amount of time that you could spend cycling becomes larger than 1,440 minutes. Explain why this study does not present findings for pollution levels lower than 24 μg/m^3. [2]

20. What other hazards of cycling in a city are not evaluated in this study? [1]

21. The research finds that in some cities, pollution is too high for people to travel to work on a bicycle and gain a health benefit from doing so. Some people might read this research and decide that it would be better to travel to work by car instead. Imagine that you work for the government of one of these cities. Write a short article that persuades people not to travel by car. [3]

◀ Aside from pollution, what other hazards may a cyclist experience traveling in a busy city? What precautions can cyclists take?

2 Balanced forces

Key concept: Systems

Related concepts: Interaction, Function

Global context: Identities and relationships

▶ The Balanced Pinnacle Rock in the Chiricahua National Monument, Arizona, appears to be about to fall over. However, the forces that act on it are balanced and the rock has been there for thousands of years and is likely to last for a long time to come. Where else do forces of nature balance?

◀ In a tug of war, the tension in the rope can be very large. Despite the large forces involved, the forces can balance so that both teams remain stationary. Only when one team can exert a little bit more force than the other team does the rope start to move one way and one team can win. In which other sporting events are opposing forces balanced, or nearly balanced?

▲ The force of gravity pulls us down towards the ground and is difficult for us to overcome. The first way in which humans were able to fly was using hot air balloons in 1783. Hot air balloons use the fact that hot air expands—when the air inside the balloon is heated, some air is pushed out. This makes the balloon lighter and able to float. As a result, the force from the air pressure is enough to balance the weight of the balloon and the people. How else have we balanced the force of gravity to enable us to fly?

◀ Balanced forces are very important in engineering and architecture because buildings have to be able to support their own weight. The Leaning Tower of Pisa is notable because at points in its history, the forces were not balanced. As a result, it began to topple. In the late 20th and early 21st centuries, efforts were made to rebalance the tower. Cables around the tower exerted a tension and hundreds of tonnes of lead were used as a counterbalance. As a result, the tower's tilt was reduced from 5.5° to under 4° and it was made safe enough to last for another couple of hundred years. Are there other examples of buildings that have required modification to help the forces to balance?

Key concept: Systems

Related concepts: Interaction, Function

Global context: Identities and relationships

Statement of inquiry:

The interaction of forces can create a balanced system.

Introduction

Forces act all around us, all the time. Whether it is the pull of gravity or the push from contact with the ground there are multiple forces acting on you at any time. In this chapter we look at different types of forces, how they act, and how we can measure them. Because forces are responsible for the interaction between different objects, one of the related concepts of the chapter is interaction.

In some situations, multiple forces act on something so that they push against each other and oppose each other. Sometimes the forces can act in such a way that they balance each other. In this chapter we will look at how we add forces and how forces can balance.

Many systems rely on maintaining a balance to function. As a result, the other related concept is function and the key concept of the chapter is systems. Sometimes the balance of these systems is not caused by forces, but other conflicting concepts that oppose each other, creating a harmonious balance.

This weightlifter needs to exert a large force to overcome the force of gravity on the weights

A governor is a device that can be used to regulate a mechanical system. The governor on this steam engine ensures that the steam engine runs at a steady speed. As the engine turns, the red spheres spin round and are flung outwards. This motion causes the sleeve at the bottom to move upwards against the force of the spring. In doing so, it lifts a lever, adjusting a valve that slows the engine down

What is a force?

A **force** is an external push or pull on an object. They are responsible for the interactions between objects. The strength of a force is measured in **newtons**.

1. Which two objects are interacting in these examples?

a) The tension in a rope when you swing on it.

b) The wind blowing on the sail of a boat.

c) The force of gravity acting on you.

What types of force are there?

One way in which objects can exert a force on each other is through contact with each other. Often these forces are called **contact forces**. When two objects come into contact a force acts outwards on each of them.

Contact forces often exert a push, but it is also possible for contact forces to pull. A rope, string or rubber band can exert a pulling force. We call this type of force a **tension**. Just as a stretched spring or rubber band will exert tension, a compressed spring will exert a **compression force**.

Another force that occurs when there is contact between two objects is **friction**. Friction acts when objects move past each other while in contact. However, friction can act even if the objects are not moving.

ABC A **force** is a push or pull on an object.

A **newton** is the unit of force.

▼ When the baseball bat makes contact with the ball, it exerts a large force. This changes the motion of the ball

ABC A **contact force** is a force that occurs between two objects in contact.

Tension is the inwards force from a stretched object such as a rope or spring.

A **compression force** is the outwards force from a compressed object.

Friction is a force that occurs when two objects are in contact and try to slide past each other.

▼ The brakes on a car use friction to exert a force. This force acts in the opposite direction to the direction of the motion, slowing the car down. There also needs to be enough friction between the car's tires and the road

▶ Friction can act even when things are not moving. These cars are parked on a steep hill. If it were not for the friction between their tires and the road, they would all slide to the bottom of the hill

Sometimes, friction is caused by objects moving through the air. This force is called **air resistance**. A similar effect can be felt by objects moving through water, however, because water is more dense, the frictional force of water resistance is greater.

▶ This car is using air resistance to slow down. The parachute at the back increases the area of the moving vehicle and therefore the air resistance increases

ABC **Air resistance** is the force felt by something moving through the air that acts against the object's motion.

Gravity is the force that acts between all objects with mass. The force of gravity is stronger when the masses are larger and the objects are closer together.

Magnetism is the force that acts between magnets. Like (same) poles will repel and unlike (different) poles attract.

How do forces act without contact?

Some forces can act between objects without any contact between them at all. These forces include **gravity** and **magnetism**. To explain how forces can act without the two objects touching, we use the idea of a field. A field is a region of space where a force may act. Around the Earth, we say that there is a gravitational field. Similarly, there is a magnetic field around a magnet.

A bar magnet is a simple type of magnet that has two poles—north and south. When it is brought near to another bar magnet, the magnetic fields of the two magnets will interact and cause a force between them. Two poles of the same type (that is, two north poles or two south poles) will repel from each other, but if the north pole of one magnet is brought near to the south pole of the other, they will attract.

Some materials are magnetic and will be attracted to either end of a bar magnet. Iron is the most common **magnetic material**, but cobalt, nickel, and some rare earth metals such as neodymium are also magnetic. Neodymium is used to make strong magnets.

ABC A **magnetic material** is a material that is attracted to a magnet. The most common magnetic materials are iron, nickel and cobalt.

Experiment

Finding the shape of a magnetic field

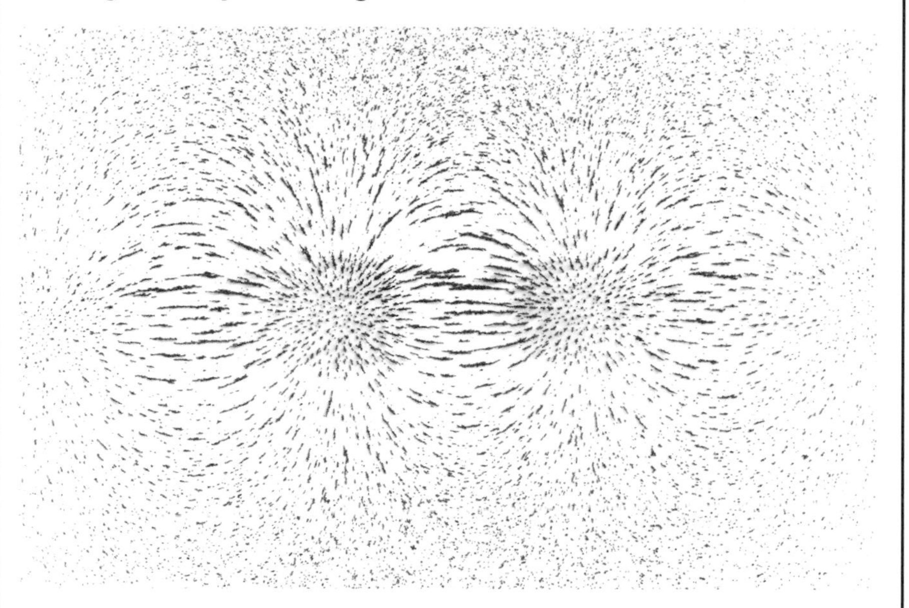

A magnetic field can be demonstrated using iron filings.

Method

- Place a piece of card over the top of a bar magnet.
- Lightly dust the card with iron filings.
- Tap the card gently to shake the iron filings into place.

The field pattern from the bar magnet shows the magnetic field from a north pole and a south pole. The field between two north poles or two south poles can be found using two bar magnets and a similar method to the one above.

Questions

1. Why are two bar magnets required to show the field between two poles of the same type?
2. Is it possible to tell which pole is which from the pattern of the iron filings?

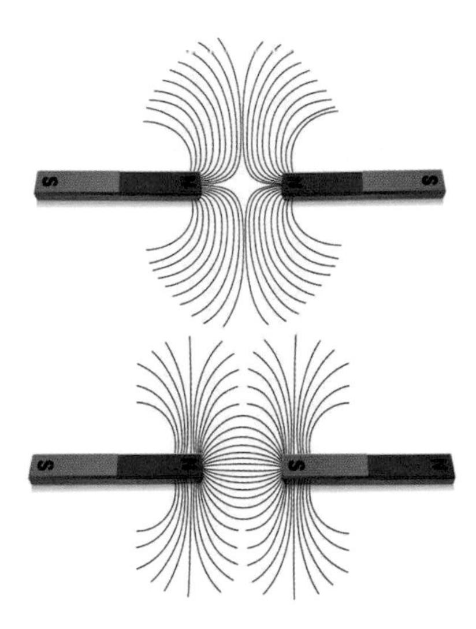

▲ The magnetic field between two magnets causes a force between them. If two north poles (or two south poles) are brought together, the magnets push away from each other. If a north and a south pole are brought together, they attract

ABC **Mass** is a property of matter that determines how an object is affected by gravity and how easy it is to accelerate it by applying a force. The mass relates to how much matter there is in the object.

Weight is the force that acts on an object due to gravity.

What is gravity?

Gravity is a force that acts between any two objects with **mass**. However, unless one of the objects is very heavy, such as a planet, it is unlikely that you would be able to detect the force of gravity. As a result, you are unlikely to ever be able to measure the force of gravity between you and someone standing next to you—this force is probably less than a millionth of a newton. However, you still feel the force of gravity that the Earth exerts on you.

Everything near the Earth feels the force of gravity pulling it towards the ground. We learn more about gravity in Chapter 4, Potential energy, kinetic energy, and gravity. The force of gravity on an object is also called its **weight**. The gravitational force, or weight, can be calculated using the equation:

$$\text{weight} = \text{mass} \times g$$

where g is the gravitational field strength. g is measured in newtons per kilogram (N/kg). On Earth, $g = 9.8$ N/kg, but it is different on other planets.

Finding your weight on other planets

Although your mass remains constant, if you went to a different planet, your weight would be different. Your weight can be calculated using the equation:

$$\text{weight} = \text{mass} \times g$$

Using the table below, calculate your weight on different moons and planets. What mass would be required to create the same weight on Earth?

Place	***g*** (N/kg)
Earth	9.8
Moon	1.6
Triton (the largest moon of Neptune)	0.78
Juno (a large asteroid in the asteroid belt)	0.14
Gliese 581d—an exoplanet that is thought to orbit a star 20 light years away (see Chapter 12, The universe, for what an exoplanet is). Astronomers believe that it orbits at a distance from its star that could give it liquid water on its surface and hence the possibility of oceans and rain.	14
Sirius B—a white dwarf star 8.6 light years away (white dwarf stars are small, very dense stars).	4,000,000

Gravitational forces act over a very long range. Here, gravity acts on Saturn's rings and on its moon, Titan, keeping them in orbit around Saturn. Because Titan is smaller than the Earth and has less mass, the gravitational field strength on Titan is much lower. On Titan, $g = 1.35$ N/kg

1. Identify the forces that are in action in these situations:

a) You are sitting on a chair

b) A carpenter sanding some wood

c) A magnet sticks to a fridge door.

How do we measure forces?

Chapter 1, Motion, looks at how direction is important when considering quantities such as position and velocity. These quantities are called **vector quantities**. Force is another example of a vector quantity because the direction in which the force acts is important, as well as its **magnitude**. For example, when a car brakes, the direction of the force is backwards and this slows the car down. The same sized force in the opposite direction would cause it to increase in speed.

When measuring the magnitude or size of a force, the **SI** unit of force is the newton, which is abbreviated to N. 1 N is approximately the force that you would need to exert to lift a 1 kg mass.

ABC

Vector quantities are quantities that have a direction and a size.

Magnitude is the size of a quantity (not taking the direction into account).

SI is an abbreviation of Système International (International System). This system of units is used by scientists for the standard units of measurement. They include the kilogram, the second and the meter.

ATL **Communication skills**

Communicating numerical quantities

When communicating the size of a quantity such as mass, speed, force or time, you need to use a unit. This helps the other person understand the quantity you are communicating. If you arranged to meet someone in 2, then they would need to know if you meant 2 minutes, 2 hours, 2 days or 2 years.

It is important that everyone uses the same units—this is why clocks have the same hours and minutes, and rulers are marked with centimeters that are the same size as every other ruler's centimeters.

To help scientists communicate quantities in a precise way, they have developed a set of well-defined units. This is called the International System (SI) and defines the meter, the kilogram and the second. Other units can be derived from these base units, such as a meter per second (m/s) for speed.

Some units have their own names, such as the newton for force, or the joule for energy. A newton is in fact one kilogram meter per second squared (kg m/s^2).

When calculating quantities and using equations, it is important to know the correct units for the numbers you are using in the calculations. This way, you can determine the units of the final answer. For example, you have seen how you can calculate speed by dividing the distance traveled by the time taken. If the units of distance are meters (m), and the units of time are seconds (s), then the units of speed will be meters per second (m/s).

$$\frac{\text{distance (m)}}{\text{time (s)}} = \text{speed (m/s)}$$

It might be necessary to convert units when calculating quantities. For example, you may need to convert units of length to meters, mass to kilograms and time to seconds.

▲ It may not look like a clock, but this is an atomic clock. This is used to help scientists define the second precisely. The clock is so accurate that it will take about 2 million years for it to lose or gain 1 second

ABC A **newtonmeter**, also called a spring balance, is a device that measures the size of a force.

A simple way of measuring forces is to use a **newtonmeter**, or spring balance. This makes use of a spring that stretches by an amount related to the size of the force acting on it.

How do forces add together?

In Chapter 1, Motion, we consider the direction of vector quantities (such as velocity) when adding them together. Two forces, each of 10 N, could add to give a force of 20 N if they both act in the same direction. If the forces acted in opposite directions, however, they would act against each other to give zero force. If the two forces acted at different angles, they would add together to give a different magnitude force, which would be somewhere between 0 and 20 N.

When two or more forces act on an object, the sum of these forces is called the **resultant force**.

ABC The **resultant force** is the sum of all the forces that act on an object.

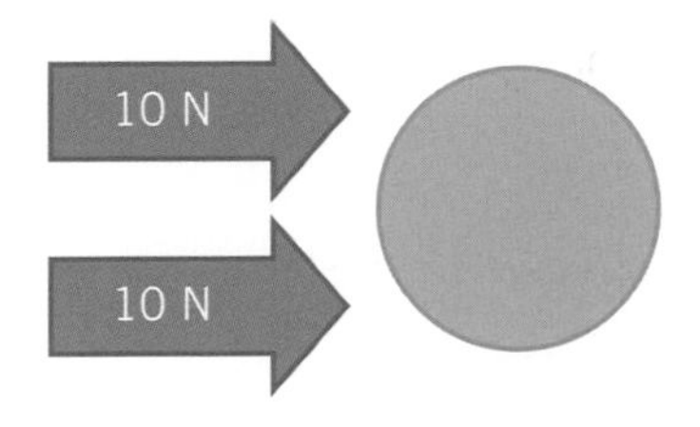

▲ These two forces add together to give a total force of 20 N

▲ Because these two forces are acting in opposite directions, the total force is zero

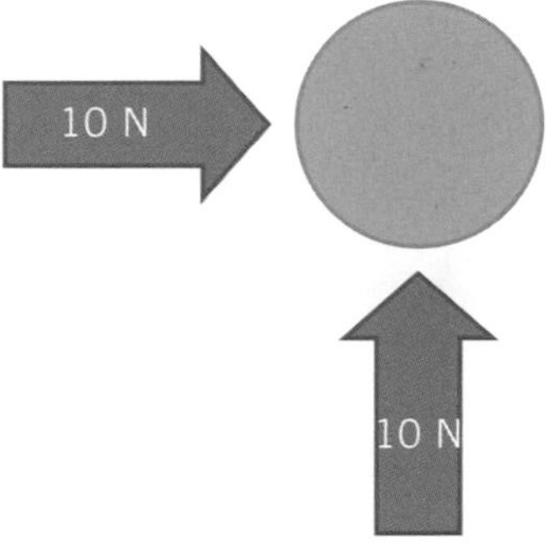

▲ These forces are acting at right angles. The resultant force can be calculated using Pythagoras's theorem to be 14 N

▼ In a canoe slalom, the canoeist has to consider the force that he exerts with his paddle as well as the force from the flowing water. These two forces may act in different directions. Paddling upstream is much harder than gliding downstream. If the canoeist wants to travel sideways across the flow of water, he must exert a force upstream with his paddle to balance the downstream force of the water

ABC When the forces acting on an object total zero, they are said to be **balanced**.

What happens when forces balance?

Sometimes two or more forces can add together to give zero force overall. When this happens, the resultant force is zero and we say that the forces are **balanced**.

In Chapter 3, Unbalanced forces, we see that an unbalanced force causes an object's motion to change. When the forces are balanced, the object's motion remains unchanged. This means that it does not accelerate or decelerate. If the object is stationary, then it remains stationary. However, if it is moving and the forces on it are balanced, the object continues to move without changing speed or direction.

▲ This Maglev train in Shanghai uses magnetic fields to exert a force that balances its weight. This means that the train is able to hover above the rails and does not need wheels. As a result, the amount of friction is reduced and the Maglev train is capable of reaching fast speeds

▲ Hummingbirds can hover next to flowers. To do this, the upward force generated by their wings must balance the force of gravity that acts downwards

▲ The Voyager 1 probe was launched in 1977. It has now got so far from the Sun, and is so distant from any other star or planet, that there are essentially no forces acting on it. Because there is no resultant force, the probe will continue to drift at its speed of 17 km/s in a straight line until it gets close enough to another star or planet that can exert a force on it

▲ The slackline walker is an example where balanced forces are important. The total force is that of the tension in the tightrope that acts on either side of the tightrope walker as well as her weight. These forces all add up to zero and the forces are balanced

What happens when objects fall?

It is important to remember that if the forces on an object are balanced, it does not necessarily mean that the object is not moving. There is a special case of forces balancing called **terminal velocity**. This can occur when something is falling. If an object is dropped from rest, the weight of the object causes it to accelerate downwards. However, as its speed increases, the force of air resistance gets larger. Eventually, the force of the air resistance has the same magnitude as the weight of the object but acting in the opposite direction. The forces are balanced and therefore the falling object no longer accelerates but keeps falling at a constant speed. This speed is called terminal velocity.

ABC **Terminal velocity** is the speed to which a falling object accelerates. At terminal velocity, the forces of weight and air resistance are balanced.

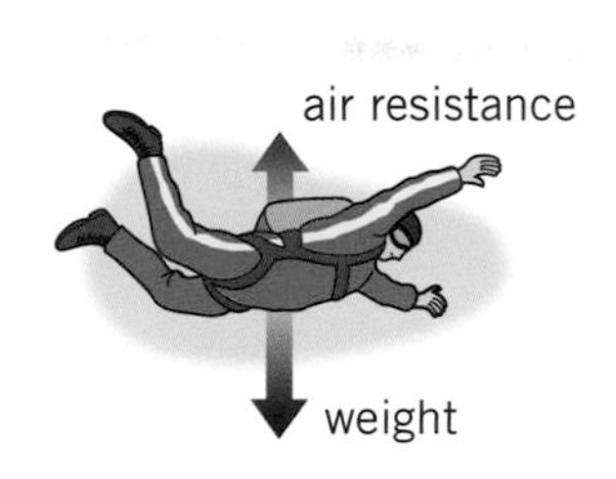

a) When the skydiver jumps out of the plane, he is falling slowly. The forces are unbalanced and so the skydiver accelerates

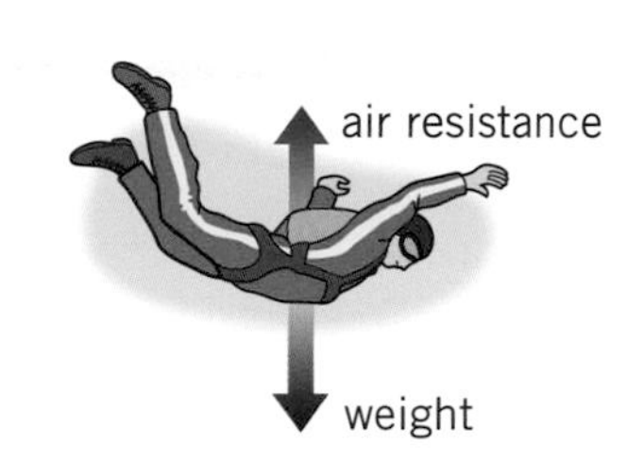

b) At terminal velocity, the forces on the skydiver are balanced. He falls at a constant speed

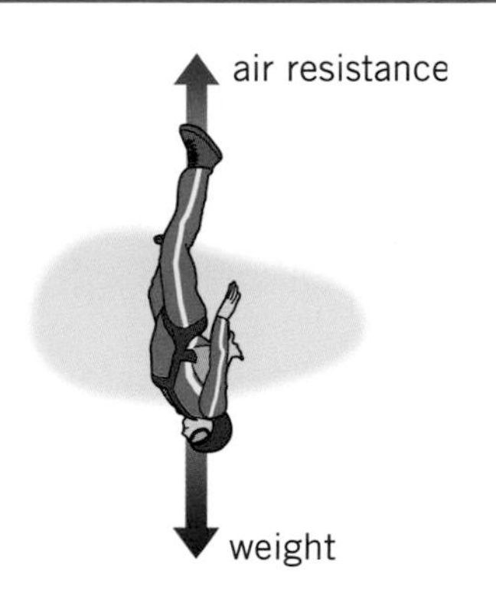

c) In a nosedive, air resistance is reduced, the forces are unbalanced and the skydiver will accelerate

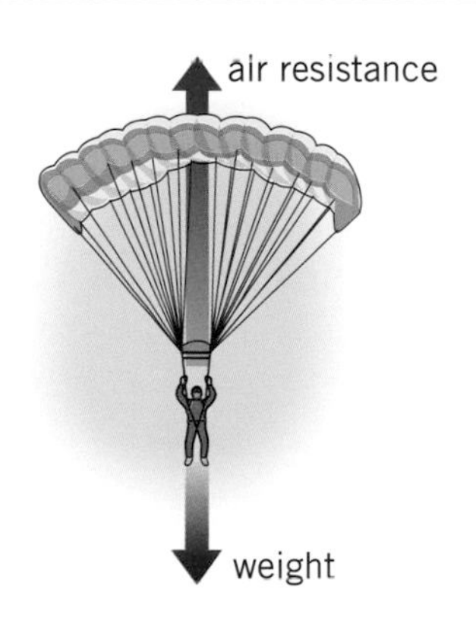

d) When the skydiver opens his parachute, the air resistance is increased. The forces are unbalanced and the skydiver decelerates

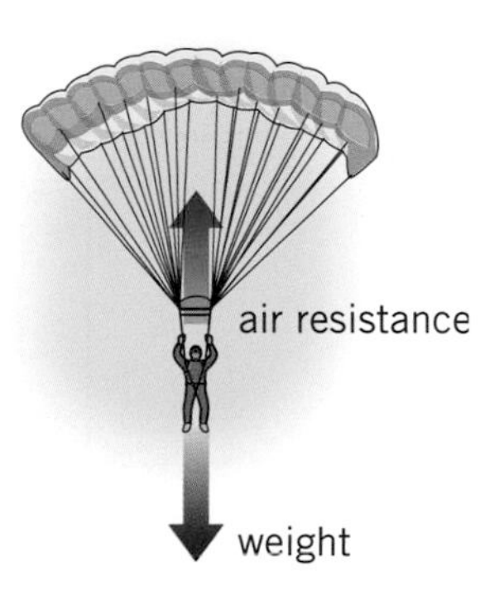

e) The skydiver is falling slower than before. The parachute provides a force that balances the skydiver's weight and so he falls at a constant speed

Skydivers use the idea of terminal velocity. If they go into a nosedive, they reduce their air resistance and they fall faster before the air resistance balances their weight. When they deploy their parachute, they increase their air resistance and therefore their terminal velocity is decreased.

Making a parachute

Design and make a simple parachute using newspaper, string, tape, and a small pebble. Your parachute should slow the fall of a stone from the height of a couple of meters.

To test the effectiveness of the parachute, drop the stone from a set height without the parachute and time how long it takes to fall to the ground. Then attach the parachute and drop the stone from the same height and time its fall again.

Find out how much longer the stone takes to fall. Compete with others to see who can make the most effective parachute.

Experiment

Finding the terminal velocity of a table tennis ball

In this experiment, you will drop a table tennis ball from increasing heights. Work in pairs so that one person drops the ball, while the other uses a stop clock to time how long it takes to fall.

Method

- Drop the table tennis ball from 0.5 m and time how long it takes to fall. Repeat this measurement three times and take an average of your results.
- Now repeat the experiment dropping the table tennis ball from 1 m. Take three measurements of the time.
- Keep repeating the experiment with different dropping heights until you get to at least 2.5 m. You may be able to use higher heights.
- Record your results in a table and plot a graph with the distance fallen on the y-axis and the time taken on the x-axis.

Questions

1. What trends do you notice in your graph?
2. Use your graph to determine the terminal velocity of the table tennis ball.

▶ To reduce the air resistance on this car, it is being tested in a wind tunnel. Its streamlined shape gives the car a lower air resistance so it can go at faster speeds

An effect similar to terminal velocity results in cars and bicycles having a top speed. As a car increases in speed, the air resistance on it increases. Cars reach a point where the engine is giving the maximum force forward, but this is balanced by the air resistance acting in the opposite direction. The car cannot go any faster than this. As a result, when building fast cars, engineers try to reduce the air resistance as much as possible by designing them with a streamlined shape.

1. Draw a diagram showing the forces that act on a car
   a) when it is stationary
   b) when it is traveling at top speed.

How fast do raindrops fall?

Scientists calculated the terminal velocity of raindrops of different sizes. They assumed that the raindrops were spherical. For raindrops of diameter 1, 2, 3, 4 and 5 mm, they calculated velocities of 2.68, 4.53, 5.89, 7.01 and 7.96 m/s.

1. Construct a table of the data.
2. Draw a graph of the scientists' results.
3. Most raindrops have a diameter of 2.5 mm. Use your graph to find the speed at which these fall.
4. In a heavy rainstorm, droplets of up to 5 mm can fall. Explain why a brief heavy rainstorm can cause more erosion than a longer period of lighter rain.
5. The scientists assumed that the raindrops were spherical in shape. Do you think this is a good assumption?

Extension: If you have already studied Chapter 4, Potential energy, kinetic energy and gravity, you could compare the kinetic energies of different raindrops. To estimate the mass of each raindrop, you must calculate the volume of a raindrop. If you treat the raindrop as a sphere the volume can be calculated using the equation:

$$\text{volume} = \frac{4}{3}\pi r^3$$

where r is the radius of the raindrop. The mass of the raindrop can be calculated using the following equation, knowing that the density of water is 1000 kg/m^3.

$$\text{mass} = \text{density} \times \text{volume}$$

Summative assessment

Statement of inquiry:

The interaction of forces can create a balanced system.

This assessment is based around the forces on cars, especially the force of air resistance and how it affects the design of cars.

Forces and cars

1. This car has broken down and is being pulled along by an SUV. The car is being pulled along a flat road at a constant speed.

Which statement correctly describes the forces?

A. The tension in the rope is larger than the car's mass.

B. The tension in the rope is equal to the car's weight.

C. The tension in the rope is equal to the contact force from the road.

D. The tension in the rope is equal to the frictional forces acting on the car.

2. When cars drive on icy roads, the friction between the car and the road is reduced. If the friction was reduced to zero, which statement describes the motion of the car along a flat icy road?

A. The lack of friction means the tires cannot grip the road and the car will stop.

B. The slippery ice results in no forces acting on the car and it will be stationary.

C. The forces on the car are balanced and it will continue at a constant speed.

D. There is an unbalanced force acting on the car and it will slow down.

3. The weight of a car is 14,400 N. What is its mass? (Assume $g = 9.8$ N/kg)

A. 147 kg

B. 1,470 kg

C. 14,400 kg

D. 141,000 kg

4. The *Curiosity* Mars rover is a car-like mobile laboratory which landed on Mars in August, 2012. The rover has a mass of 900 kg. On Mars, $g = 3.72$ N/kg. Calculate the weight of *Curiosity* on Mars.

A. 242 N

B. 413 N

C. 3348 N

D. 8820 N

5. The diagram shows some forces acting on a car. If the forces in the diagram are balanced, what is the value of the missing force?

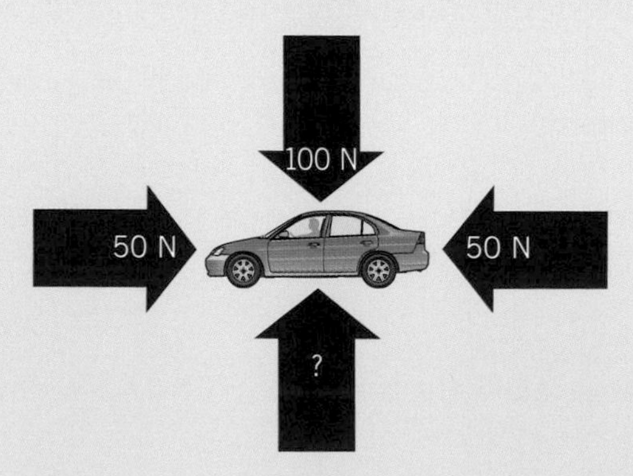

A. 0 N

B. 50 N

C. 100 N

D. 200 N

Investigating air resistance in a wind tunnel

Engineers use wind-tunnels to investigate the air resistance on cars. As part of the process to design a new car, engineers put a model car in a wind tunnel and vary the wind speed through the tunnel. They measure the air resistance that acts on the model car.

Wind speed (m/s)	Force (N)
0	0
1	0.03
2	0.13
3	0.30
4	0.54
5	0.84

6. Identify the independent and dependent variables in this experiment. [2]

7. Suggest a suitable control variable for the experiment. [1]

8. What are the advantages and disadvantages of using a model car in a wind tunnel rather than using the real car? [3]

9. Plot a graph of the data in the table on the right and add a line of best fit. [4]

Designing an aerodynamic car

The designers of a car investigate its aerodynamic drag. The more streamlined it is, the lower the value of its aerodynamic drag. The graph shows the top speed of their car for different amounts of aerodynamic drag and for different forces that the engine provides.

Their car's engine currently produces a force of 1,200 N and the car has a top speed of 40 m/s.

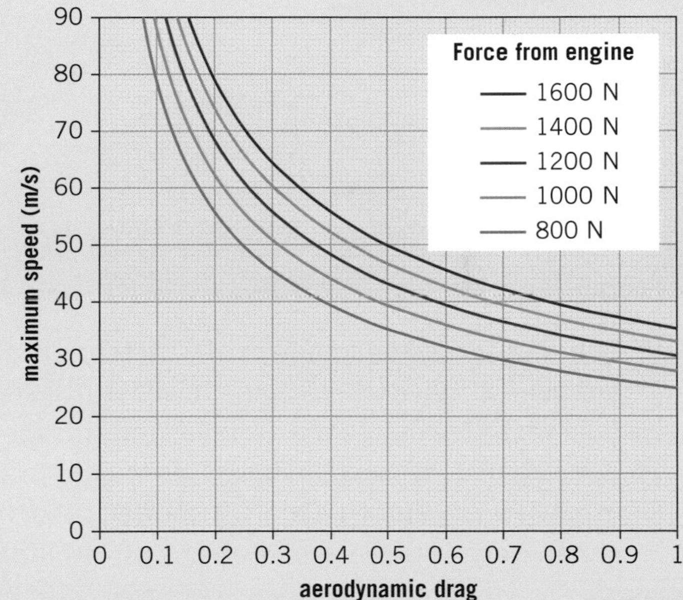

10. At its top speed, the air resistance acting on the car balances the force from the engine. What is the strength of the air resistance that acts on the car? [1]

11. Name one other force that acts on the car. [1]

12. Use the graph to estimate the aerodynamic drag of the car. [2]

13. If they improved the aerodynamic drag by 0.1, what force would be necessary to keep the top speed at 40 m/s? [3]

14. The designers investigate the costs of making improvements to their car. Cutting the aerodynamic drag by 0.2 will add $1,000 to the cost of the car. Putting in an engine that can provide a driving force of 1,600 N would cost the same amount. Which of these improvements would give the greatest increase in top speed? [5]

15. Designing a car so that it has less aerodynamic drag makes it more fuel efficient. Explain two advantages of cars that are more fuel efficient. [3]

16. The designers of the car debate whether they should publish their experimental data. Write one argument for sharing the data, one argument against publishing it, and form a conclusion. [5]

3 Unbalanced forces

Key concept: Relationships

Related concepts: Energy, Consequences

Global context: Scientific and technical innovation

Caenorhabditis elegans is a nematode worm that lives in the soil. It is small—about 1 mm in length—but has been found to be able to withstand huge accelerations (up to 400,000 times the freefall acceleration on Earth). The force required to create this acceleration is only about 0.03 N, because of the small size of the worm. To achieve a similar acceleration in a cow would take a force of about 3 billion newtons. Their ability to survive large forces and accelerations makes them very tough and adaptable to hostile conditions. The worms have been taken to the International Space Station, and some worms were even discovered to have survived when the Space Shuttle *Columbia* broke up on re-entry in 2003. How can studying small animals lead us to discover more about ourselves?

This long-exposure image shows meteors, or shooting stars. Every August, the Earth's orbit passes through a region of space where Comet Swift–Tuttle has left a trail of dust. The Earth's gravity exerts a force on the dust and accelerates the particles towards the Earth. When they enter the Earth's atmosphere, the work done against air resistance causes energy to be transferred to heat. The comet fragments burn up in the atmosphere, causing a meteor shower. Other meteor showers occur at other times of the year. How else can material from space interact with our atmosphere?

► For a rocket to accelerate off the ground, it must generate an unbalanced force. Because the weight of its fuel is so great, the amount of upward force required is huge. The Saturn V rocket, which was used for the Apollo moon landings, had a weight of about 30 million newtons and a thrust of 35 million newtons. A great deal of research and innovation was required in order to find suitable fuels that could generate a large force without being too heavy. What other technical innovations were driven by the exploration of space?

▼ Yellowstone National Park is situated on a volcanically active site and has many geysers. A geyser occurs when water seeps downwards through the rock surface and reaches hot volcanic rock. This water boils and pressure builds up on the water above. Eventually, the pressure is big enough that there is an unbalanced force acting upwards on the water and it is pushed upwards out of the geyser. This picture is of Castle Geyser, which erupts in this way for about 20 minutes every 10 to 12 hours. The unbalanced force on the water is enough for the geyser to reach a height of about 27 m. How else can the Earth exert a force on things?

Statement of inquiry:

The relationship between unbalanced forces and energy has enabled huge improvements in technology.

Key concept: Relationships

Related concepts: Energy, Consequences

Global context: Scientific and technical innovation

Statement of inquiry:

The relationship between unbalanced forces and energy has enabled huge improvements in technology.

Introduction

In Chapter 2, Balanced forces, we look at how balanced forces behave. However, forces are not always balanced. Sometimes objects have forces acting on them that do not add to zero. These unbalanced forces cause things to move in a certain way. In this chapter, we will see how unbalanced forces change an object's motion. As a result, one of the related concepts for this chapter is consequences.

Unbalanced forces cause objects to accelerate. This means that their velocity changes and therefore the object may gain or lose energy. In this chapter, we see how energy and forces are related. The key concept of the chapter is relationships and the second related concept of the chapter is energy.

Many of the discoveries of how forces cause things to move were made by Sir Isaac Newton in the 17th century. His laws of motion are often regarded as one of the most important stages in the development of physics. Understanding how forces and energy behave is important in order for us to understand how the universe works. It has enabled us to build machines and improve on them. For this reason, the global context of the chapter is scientific and technical innovation.

▲ These young Maasai men in Kenya are using the friction between two sticks to start a fire. By doing work against this friction force, they can generate heat energy. The use of fire was a significant innovation in human history as it enabled us to keep warm and cook food. Later, during the industrial revolution, we were able to convert some of this heat energy into useful mechanical work

▲ The ThrustSSC (supersonic car) holds the land speed record of just under 1228 km/h. To travel so quickly, the car had to be designed to generate a large unbalanced force. Its jet engines produced a force of over 220,000 N, and it was designed to have as little air resistance as possible

What is an unbalanced force?

In Chapter 2, Balanced forces, we saw how forces add together and how they can end up being balanced. This happens when the sum of all the forces acting on an object is zero, and as a result there is no net force. Sometimes the forces do not balance, and the net force is not zero. If this is the case, then the net force is called an **unbalanced force**.

ABC An **unbalanced force** occurs when the sum of the forces that act on an object is not zero.

Remember that the direction of forces is important. If two forces both act in the same direction, their magnitudes are added together. If the forces act in opposite directions, then their magnitudes are subtracted to find the overall force. If two forces act at right angles to each other, the magnitude of the resultant force can be found using Pythagoras's theorem. For example, imagine an object with a weight of 4 N is dropped. The wind blows on the side of the object, exerting a force of 3 N. The total force that acts can be calculated using:

$$F^2 = 3^2 + 4^2$$

$$F^2 = 25$$

$$F = 5\text{ N}$$

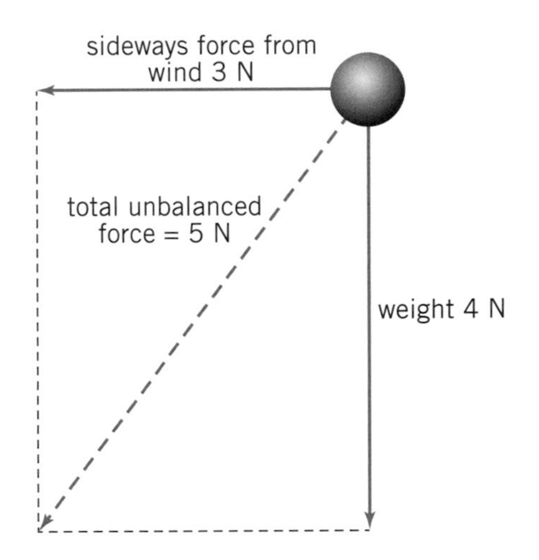

1. For each of these diagrams, calculate the unbalanced force on the basketball. Remember to include the direction.

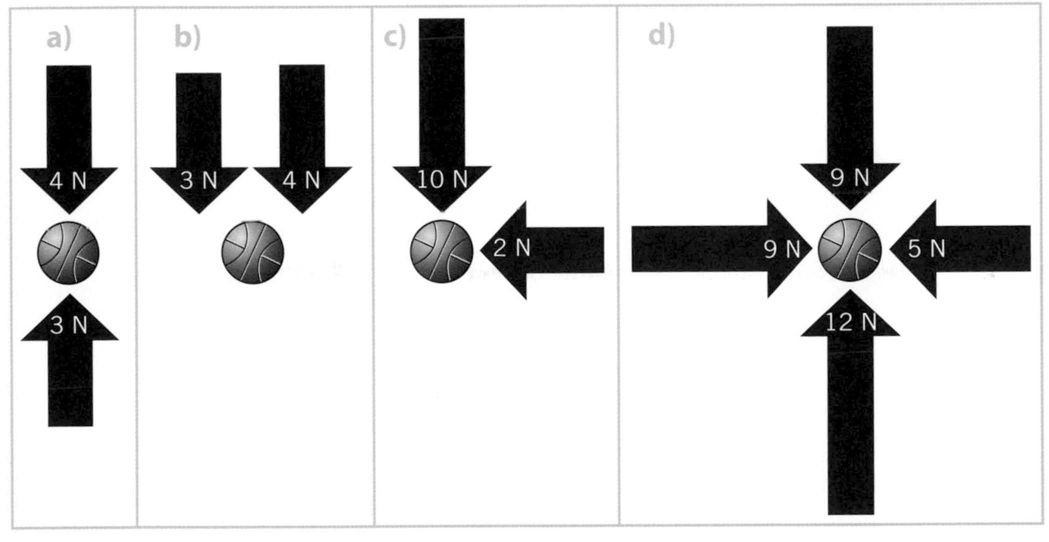

▲ In this stunt, an escapologist hangs from a burning rope. Before the rope burns through, the tension in the rope balances the weight of the stunt man. When the rope burns through and breaks, the forces on the escapologist are unbalanced and he will accelerate to the ground. Of course, he has to escape before that happens

What do unbalanced forces do?

Chapter 2, Balanced forces, looks at how the motion of an object is unchanged when the forces on it are balanced. This means that a stationary object will remain stationary, and a moving object will continue to move with the same velocity—its speed and direction are unchanged.

When an unbalanced force acts on an object, its velocity changes. This may mean that its speed or direction (or both) change and it will accelerate (see Chapter 1, Motion, for the relationship between velocity and acceleration). As a result, we say that unbalanced forces cause acceleration.

ABC **Newton's second law** is a law of motion formulated by Isaac Newton that may be expressed as the equation force = mass × acceleration.

Two variables are said to be **proportional** if increasing one means that the other increases as well.

The relationship between the unbalanced force and the acceleration is:

$$\text{force } (F) = \text{mass } (m) \times \text{acceleration } (a)$$

This is often referred to as **Newton's second law**. It means that an object with a large mass requires more force to accelerate it than a small mass. It also means that if the force acting on an object increases, the acceleration increases. This means force and acceleration are **proportional**.

This scooter is fully loaded with packages. Due to its increased mass, it is slower to accelerate and also harder to stop. Even turning corners will be more difficult

ATL Transfer skills

Isaac Newton

In the 17th century, British physicist and mathematician Isaac Newton considered the motion of planets. It is reported that he saw an apple falling from a tree and wondered why it fell straight down rather than sideways. He realized that there must be a force that acted from the center of the Earth on the apple.

Newton started to extend his thinking beyond the apple. He wondered how far this force extended and he concluded that the same force—gravity—was responsible for keeping the Moon in orbit around the Earth.

The result of Newton's work was a book called *Philosophiæ Naturalis Principia Mathematica* (at that time Latin was considered the language of scholarship; the book title means the *mathematical principles of natural philosophy*). In it, he linked his observations of what had happened to the apple with his thoughts on gravity and used much of his previous mathematical work. In this work, Newton presented three laws of motion that describe balanced and unbalanced forces.

By transferring these ideas from one situation to another, Newton was able to calculate the motion of planets about the Sun, as well as the motion of objects on Earth that experience forces.

Transferring ideas from one topic to another has often been important in making scientific progress. Today, many researchers work across multiple different kinds of science such as physical sciences, life sciences and computer science. Together, they tackle topics that include studying genomes, artificial intelligence and the possibility of finding life forms on other planets.

What are the benefits of scientists working in teams?

Investigating Newton's second law

The picture shows a way of investigating Newton's second law. A trolley was put on a ramp and a newtonmeter was used to measure the unbalanced force down the slope. The trolley was then released and the time it took to roll to the bottom of the slope was measured. The experiment was repeated three times. The height and angle of the ramp was increased so that the trolley experienced a greater force down the slope and the experiment repeated again.

The distance to the bottom of the slope was 0.94 m.

1. The table below shows the results that were obtained. Complete the column for average time. The first one has been done for you here. Alternatively, you can use the method in the experiment on page 44 to produce your own data.

Force (N)	Time (s)			Average time (s)	Acceleration (m/s^2)
	1	2	3		
1.0	2.01	2.12	2.10	2.08	
1.5	1.57	1.69	1.66		
2.0	1.31	1.33	1.33		
2.5	1.15	1.17	1.21		
3.0	1.03	1.05	1.02		
3.5	0.96	0.97	1.03		
4.0	0.90	0.90	0.88		

2. The acceleration can be found using the formula:

$$x = \frac{1}{2}at^2$$

where x is the distance down the slope, a is the acceleration, and t is the time that it takes if it starts from rest. Rearranging this formula gives:

$$a = \frac{2x}{t^2}$$

which can be used to find the acceleration.

For example, here is how you calculate the acceleration using the average time above:

$$a = \frac{2 \times 0.94}{2.08^2} = 0.435 \text{ m/s}^2$$

Calculate the acceleration for the other average times.

3. Plot a graph of force against acceleration.

4. Newton's second law is $F = ma$. Does your graph support Newton's second law?

Experiment

Investigating the link between force and acceleration

For this experiment you will need a ramp that is approximately 1 meter long and a trolley to roll down it. You will need a ruler to measure the length of the ramp, a stop clock and a newtonmeter (see photo on previous page).

Method

- Tilt the ramp at a small angle. Measure the distance that the trolley travels from the start of the ramp to the end.
- Place the trolley on the ramp. If possible add masses to the trolley so that the trolley has a mass of about 2 kg.
- Use a newtonmeter to measure the unbalanced force on the trolley down the slope.
- Use the stop clock to measure the time it takes for the trolley to roll down the ramp. Ensure that you catch it at the end. Repeat this three times and calculate an average time.
- Repeat the experiment with the ramp tilted at different angles. Each time measure the unbalanced force on the trolley and the time taken for the trolley to roll down the ramp.
- Construct a table of your data:

Unbalanced force (N)	**Time (s)**	**Final speed (m/s)**	**Acceleration (m/s²)**

Questions

1. To calculate the final speed, use the equation:

$$\text{final speed} = \frac{2 \times \text{distance}}{\text{time taken}}$$

The reason for the 2 in this equation is because we want the final speed. Remember the equation for average speed from Chapter 4, Motion:

$$\text{average speed} = \frac{\text{distance}}{\text{time taken}}$$

Because the trolley is accelerating, it is below the average speed for the first half of the time and above the average speed for the other half of the time it takes the trolley to roll down the slope. The average speed is therefore half of the final speed.

2. To calculate the acceleration, use the equation:

$$\text{acceleration} = \frac{\text{final speed}}{\text{time taken}}$$

3. Now plot a graph of acceleration against unbalanced force. Does your graph support Newton's law that force is proportional to acceleration?

The first vehicle to travel at 100 mph

In 1904, Charles Rous-Marten traveled on the *City of Truro* locomotive. He timed part of the journey, which showed that it had become the first vehicle ever to go faster than 100 miles per hour. He later wrote:

"On one occasion when special experimental tests were being made with an engine [...] hauling a load of approximately 150 tons [...] down a gradient of 1 in 90, I personally recorded a rate of no less than 102.3 miles an hour for a single quarter-mile, which was covered in 8.8 seconds...; five successive quarter-miles were run respectively in 10 seconds, 9.8 seconds, 9.4 seconds, 9.2 seconds and 8.8 seconds. This I have reason to believe to be the highest railway speed ever authentically recorded."

Charles Rous-Marten, p2118, *Bulletin of the International Railway Congress*, October 1905

(NB Before the metric tonne, which is 1000 kg, was used, the ton had different meanings and definitions in different countries.)

The *City of Truro* locomotive

1. A quarter-mile is 402 m. Use the information in the text to complete the table of distance against time.

Total distance traveled (m)	Total journey time (s)
0	0
402	10
804	19.8

2. Plot a graph of your values of distance (*y*-axis) and time (*x*-axis). How is the train's acceleration shown in the graph?
3. The first 402 m were covered in 10 s and the final 402 m were covered in 8.8 s. Calculate the initial and final speeds as the *City of Truro* traveled down the slope.
4. If you assume that the speed is taken at the end of each time interval, it takes 37.2 s ($9.8 + 9.4 + 9.2 + 8.8 = 37.2$) for the engine to accelerate from its initial to final speed. Use your values of the initial and final speeds to calculate the acceleration of the engine.
5. The mass of the trucks is reported as being at least 150 tons, which is equivalent to 152,400 kg. Calculate the force that the engine exerted on the trucks.
6. The speed record set by *City of Truro* is disputed and is often discounted as it was not independently verified. Suggest one way in which you might improve the speed measurement that Rous-Martin described.

1. A sprinter has a mass of 80 kg and can accelerate at 4 m/s^2 at the start of a race. Calculate the unbalanced force on the sprinter.
2. A train has a driving force of 160,000 N and a mass of 800,000 kg. Calculate its acceleration.
3. A motorbike with a mass of 300 kg (including the rider) accelerates from 0 to 30 m/s in 8 s. Calculate the force required to do this.

What is work?

An unbalanced force may act to accelerate or decelerate an object. A force may also be applied to move an object against an existing force. This could be lifting an object against the force of gravity, compressing a spring, or even pushing something across a rough surface against friction. In all of these cases, the force has transferred energy and we say that work has been done. There is more about energy in Chapter 4, Potential energy, kinetic energy and gravity.

ABC

Work is the energy transferred when a force is applied.

The **joule** is the SI unit of energy and work.

When a force accelerates or decelerates an object, the kinetic energy it has is changed. For example, when an object is lifted against the force of gravity, its gravitational potential energy is increased. When a spring is compressed, then energy is stored as elastic potential energy. When work is done against friction, energy is transferred to heat energy of the object and the surface. The amount of energy transferred is called the **work** done by the force. Because energy is measured in **joules** (J), the unit of work is also joules.

▲ When the archer pulls the bow back, she does work in stretching the bow. Energy is transferred into elastic potential energy of the stretched bow. When the archer releases the bow, there is an unbalanced force acting on the arrow. Work is done on the arrow and it accelerates, gaining kinetic energy

The amount of work done, or energy transferred, can be calculated using the equation:

$$\text{work done} = \text{force} \times \text{distance}$$

where the distance is the distance over which the force acts. The distance should be measured in meters and the force should be in newtons. Using these units will give the work done in joules (J).

For example, imagine an archer pulling the string of a bow back by 60 cm. This requires a force of 100 N. How much work does the archer do?

First, you need to convert the distance into meters: 60 cm = 0.6 m

Then use the equation work done = force × distance

$$= 100 \times 0.6$$

$$= 60\ \text{J}$$

1. A person pushes a car. He exerts a force of 800 N over a distance of 10 m.

   a) Calculate the work done on the car.

   b) The car and its contents have a mass of 1,200 kg. Calculate its acceleration.

   c) A passenger gets out of the car and helps to push. Describe the ways in which this changes the work done on the car and the acceleration.

   d) If the passenger exerts the same 800 N force and has a mass of 70 kg, calculate the work done over 10 m and the new acceleration of the car.

Summative assessment

Statement of inquiry:

The relationship between unbalanced forces and energy has enabled huge improvements in technology.

As our knowledge of forces and energy has improved and our technical and engineering ability has increased, we have been able to improve the transport systems that we use. This assessment is themed around the forces that are involved in transport and how they may develop in the future.

Unbalanced forces and transportation

1. A motorbike accelerates with a force of 1,500 N. It experiences air resistance, which acts in the opposite direction with a magnitude of 1,200 N. Calculate the unbalanced force that acts on the motorbike.

   A. 150 N
   B. 300 N
   C. 1,500 N
   D. 2,700 N

2. A horse pulls a cart that requires a force of 200 N. How much work is done in pulling the cart 250 m?

   A. 8,000 J
   B. 45,000 J
   C. 50,000 J
   D. 125,000 J

3. A car rolls down a hill with an acceleration of 1.5 m/s^2. If the mass of the car is 900 kg, calculate the unbalanced force that acts on the car.

   A. 150 N
   B. 600 N
   C. 900 N
   D. 1,350 N

4. When an unbalanced force acts on a vehicle, which of these quantities changes?

   A. Acceleration
   B. Velocity
   C. Speed
   D. Energy

5. A train pulls some carriages. The train exerts a force of 40,000 N and frictional forces act in the opposite direction with a force of 35,000 N. The total mass of the train and carriages is 200,000 kg. What is the acceleration of the train?

   A. 0.025 m/s^2
   B. 0.175 m/s^2
   C. 0.2 m/s^2
   D. 0.375 m/s^2

Testing different fuels

When designing steam engines as well as modern engines, it has been important to consider which fuel to use, as a fuel that releases more energy can do more work.

In a test to determine how much energy is stored by different fuels, 10 g of each fuel was burned to heat some water. The amount by which the temperature of the water rose can be used to calculate the amount of energy released.

6. Determine the independent variable of this investigation. [1]

7. Identify one control variable. [1]

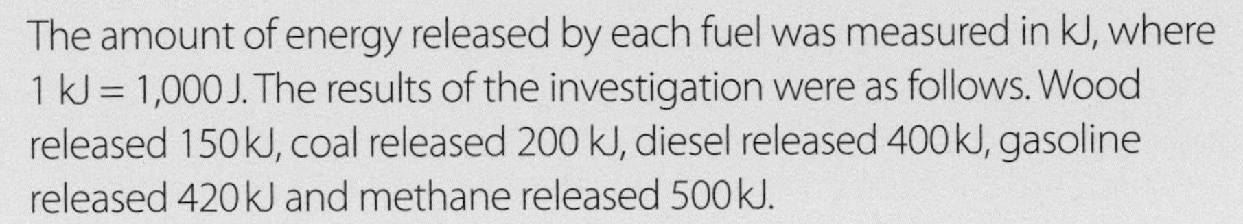

The amount of energy released by each fuel was measured in kJ, where 1 kJ = 1,000 J. The results of the investigation were as follows. Wood released 150 kJ, coal released 200 kJ, diesel released 400 kJ, gasoline released 420 kJ and methane released 500 kJ.

8. Present the data in a suitable table. [2]
9. Explain why a bar chart is more suitable than a scatter graph to present this information. [2]
10. Draw a bar chart to represent the findings of the experiment. [4]

Electric cars

Improvement in battery technology has enabled electric cars to become a viable form of transport.

The battery of an electric car needs to be charged in order to provide the energy to do work

The graph below shows the amount of work (in megajoules or MJ) that an electric car does when traveling a distance of 1 km, at different speeds. (1 MJ = 1,000,000 J)

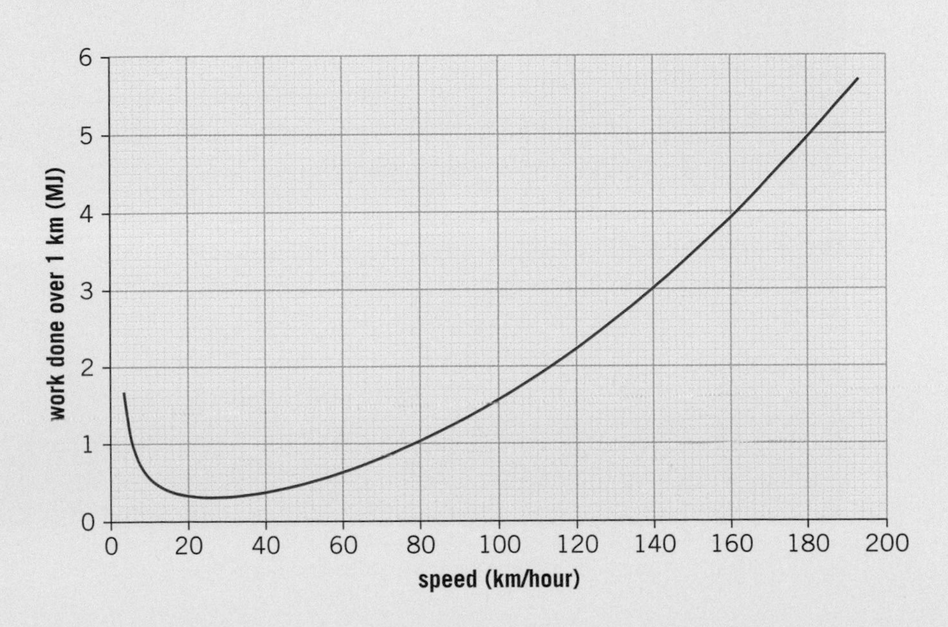

11. Suggest why the car needs to do more work at higher speeds. [2]

12. At what speed should the car travel for it to be most efficient? [1]

13. The car has a maximum acceleration of 4 m/s^2. The car's mass is 1,250 kg.

a) Calculate the maximum force that the car can exert. [1]

b) Calculate the maximum work that the car could do over the distance of 1 km. [1]

c) Hence use the graph to find the car's maximum speed. [1]

A driver wants to travel 250 km without recharging. The battery in the electric car stores 250 MJ.

14. Calculate the average amount of work per kilometer that the car can do over this journey without recharging, and hence find the shortest time in which the driver could complete the journey. [4]

Using solar sails to propel spacecraft

In the future it might be possible, or even necessary, to use spacecraft to travel to nearby stars. Solar sails are a proposed way of propelling spacecraft. They would work by constructing a reflective sail that has a large area and is very light. As the Sun's light bounces off the sail, it would exert a small force.

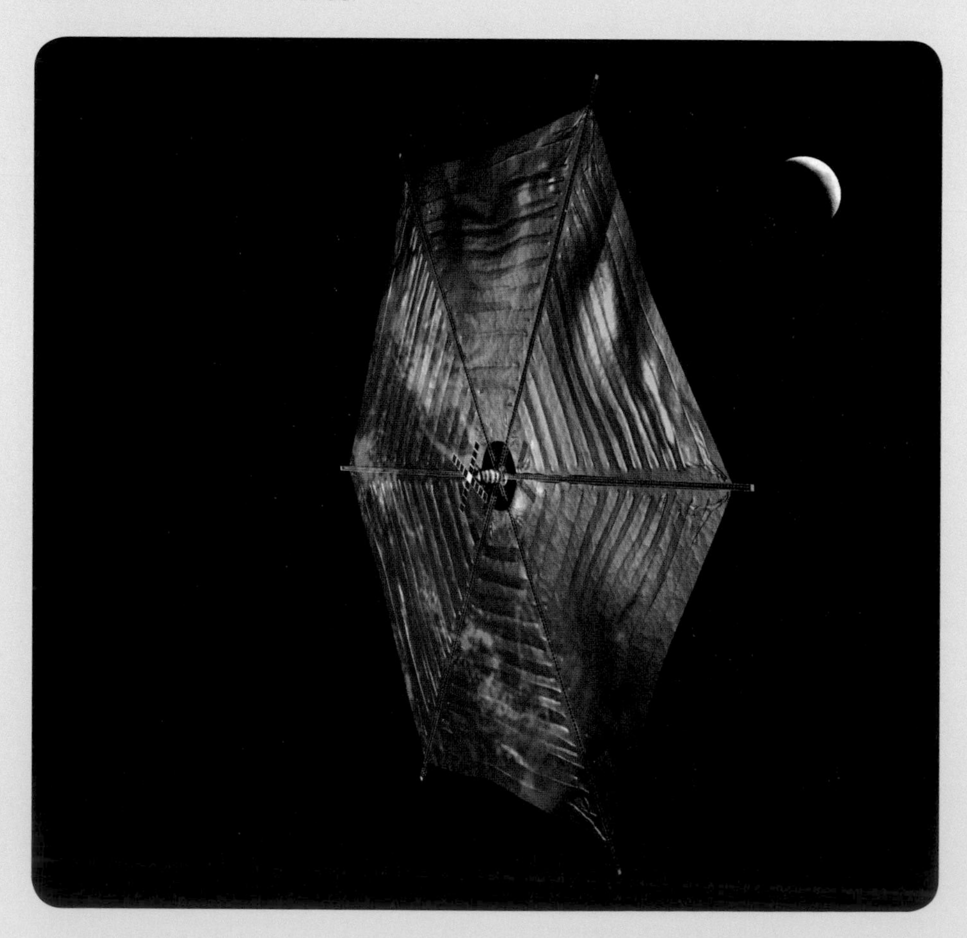

An artist's impression of what a solar sail would look like in space

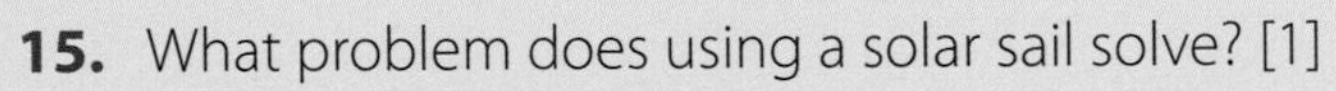

15. What problem does using a solar sail solve? [1]

16. Explain one advantage and one disadvantage of using a solar sail over a normal rocket. [2]

17. The force on the sail would be small, approximately 8 millionths of a newton per square meter (N/m^2). If a square sail were made with a side of length 100 m, show that the force on the sail is just less than a tenth of a Newton. [2]

18. If this force were to pull a spacecraft with a mass of 20,000 kg, calculate the acceleration of the spacecraft. [2]

19. If the force and acceleration remain constant, calculate the speed of the spacecraft after:

a) 1 day

b) 1 year. [3]

20. The distance s that is traveled under constant acceleration is given by the equation:

$$s = \frac{1}{2}at^2$$

where a is the acceleration and t is the time in seconds. Use this equation to calculate how far the spacecraft would travel in 1 year. [3]

4 Potential energy, kinetic energy and gravity

Key concept: Change

Related concepts: Form, Energy

Global context: Orientation in space and time

◀ Getting to the planet Mercury is difficult. The Sun's gravitational force would accelerate a spacecraft towards it so that by the time it reached Mercury's orbit it would travel past very quickly. More fuel is required to slow down and land on Mercury's surface than would be required to escape the Earth's gravity. What are the energy transfers involved in sending a space probe to another planet?

◀ This ski jumper generates a large upwards velocity. Their kinetic energy is converted into potential energy and they are able to gain a large height. What are the energy changes involved with other events or sports?

▶ In 1960, Joseph Kittinger went up in a balloon to an altitude of over 31 km. He then jumped out and performed the highest ever skydive. He was in free-fall for 4 minutes 36 seconds during which time his gravitational potential energy was converted into kinetic energy. He reached a speed of 988 km/h. Later, at the age of 84, he directed Felix Baumgartner as he broke Kittinger's record for the highest skydive in 2012, with a height of 39 km. Which other people have made pioneering missions into space?

Statement of inquiry:

Changes in energy drive the basic processes of nature.

▶ A neutron star is a very strange object indeed. Their densities are predicted to be about 5×10^{17} kg/m^3. This means that 1 cm^3 of material would have a mass of 500 million million tonnes. The large density gives neutron stars a very large gravitational field. Your weight on a neutron star would be about 200 thousand million times what it is on Earth. The nearest known neutron star is over 400 light years away. How can scientists determine properties such as the density of a neutron star?

Key concept: Change

Related concepts: Form, Energy

Global context: Orientation in space and time

Statement of inquiry:

Changes in energy drive the basic processes of nature.

▼ This rock is a fragment of the Nakhla meteorite. It is named after the village of El Nakhla El Bahariya, Egypt, where it fell to Earth in 1911. This rock started out on Mars. An asteroid impact threw it off the surface of Mars about 10 million years ago with a high enough velocity that the force of Mars's gravity was not able to do enough work to stop it escaping. As it fell from space, the force of gravity did work on the meteorite accelerating it towards the ground. Some of the work was done against air resistance and the meteorite would have become very hot before it exploded above the Earth. Over 200 meteorites from Mars have been found on Earth

Introduction

In Chapter 3, Unbalanced forces, we introduced the concept of energy and work. When we do work against a force, it requires energy. Energy is one of the most fundamental and important ideas in physics and so one of the related concepts for this chapter is energy.

Energy is important because it allows us to do things. It gives things the capacity to change or move and as it does so, the energy can be changed from one form into another. The fundamental processes of nature are driven by these transfers of energy. Falling objects convert gravitational potential energy into kinetic energy. Moving things lose energy through friction and so slow down. When objects scrape against each other they transfer energy to heat and so get hot. As a result, the key concept of this chapter is change and the other related concept is form.

In this chapter we will look at two important forms of energy: kinetic energy, which is possessed by moving objects, and potential energy, which is a stored form of energy. We will see some of the ways in which energy can be stored as potential energy, including gravitational potential energy, which can be given to an object by lifting it.

▲ The sport of mountain running (also called fell running, hill running or trail running) involves running on steep hills. Running uphill requires the runner to do work against gravity and so he needs more energy. When the runner goes downhill, gravity helps him to convert gravitational potential energy into kinetic energy, making it easier

▲ The Nakhla meteorite is estimated to have been 10 kg before it broke up. It caused very little damage on Earth. The meteorite that caused this crater in Arizona is thought to have been 50 m across. When it struck, around 50,000 years ago, it is thought to have had an energy of about 40,000 million million (4×10^{16}) J

What is energy?

Energy is one of the most fundamental concepts in physics. It is the property that allows work to be done on an object (see Chapter 3, Unbalanced forces, for how work is done when a force moves an object). This might be pushing against a force such as gravity in order to lift an object, or it might be a force accelerating an object when it falls. Giving energy to an object can also causes it to heat up.

The **law of the conservation of energy** is fundamental to understanding energy. It states that energy cannot be created or destroyed, but may change form. This means that energy may be transferred from one thing to another, and it may take on different forms, but the total amount of energy must remain the same.

ABC The **law of the conservation of energy** states that energy cannot be created or destroyed. Instead, it is transferred between objects and changed into different forms.

ATL **Critical thinking skills**

Observing carefully in order to recognize problems

The idea of perpetual motion—motion that would keep going indefinitely without the need of a power source—has intrigued scientists for many centuries. On the right is a model based on a diagram by Leonardo da Vinci. The ball bearings supposedly roll towards the rim of the circle on the left-hand side of the wheel and so can turn the wheel, more easily than the ball bearings on the right-hand side of the wheel, which roll towards the pivot. As a result, it is suggested that the wheel would turn anticlockwise indefinitely. If this were to happen, then the law of the conservation of energy would be broken since the wheel would generate its own kinetic energy!

Da Vinci appreciated that this machine (and others that he designed) could never work. Although the ball bearings on the left can exert a larger turning force, there are more ball bearings on the right-hand side.

Find the flaw in this description of a machine:

"A solar panel is used to convert the Sun's light into electricity. Some of the electricity that is generated powers a strong lamp that also shines on the solar panel. This boosts the amount of light that hits the solar panel, and therefore more electrical power is generated, which means that some can be spared to power the lamp. If the lamp is made bright enough (or other lamps are added), the solar panel does not even need the Sun's light to operate."

Units of energy

As with many physical quantities, there are many different units that are used to measure energy. The SI unit used by scientists is the joule (J). Using the equation:

work done (J) = force (N) × distance (m) (see Chapter 3, Unbalanced forces)

then if a force of 1 N is applied over a distance of 1 m, the energy transferred is 1 J.

The table shows some other units of energy that are sometimes used.

Unit	Uses	Equivalent in joules
foot-pound	used in a system when distances are measured in feet and masses in pounds	1.36
kilocalorie (kcal)	old unit of energy that is still used to give the nutritional energy contained in food	4200
tonne of TNT	The energy released when a tonne of TNT is exploded. It is used to compare explosives	4.184 billion (4.184×10^9)
cubic mile of oil	The amount of energy released by burning a cubic mile of oil. Used (rarely) to communicate energy use on a large scale.	160 million million million (1.6×10^{20})

1. An apple will supply about 90 kilocalories when you eat it. Convert this energy to:
   a) joules
   b) foot-pounds.
2. The total amount of explosives used in World War II was equivalent to about 3,000,000 tonnes of TNT. Convert this energy to:
   a) joules
   b) cubic miles of oil.
3. In 2016, the world's consumption of energy resources was approximately equivalent to 4 cubic miles of oil. Convert this value to:
   a) joules
   b) tonnes of TNT.

Potential energy is a form of energy stored by doing work against a force.

Chemical potential energy is the form of energy stored in chemical elements and compounds. This energy may be released during exothermic chemical reactions, or taken in and stored during endothermic reactions.

What forms can energy take?

Energy can be stored in many forms and can be transferred between these forms. Sometimes it is stored in the form of **potential energy**, waiting to be released. At other times the energy may be transferred to a different object or converted into a different form.

In Chapter 9, Chemical reactions, we see how chemical reactions can cause temperature changes. This shows us that chemicals can store energy and that in a chemical reaction, when one chemical substance changes to another, the amount of energy that is stored changes. If the result is that more energy is stored, then this energy will have to have come from the surroundings—this reaction will be endothermic. If the amount of energy stored decreases, then the excess energy is released to the surroundings and the reaction will be exothermic.

▲ A chemical icepack can be used to treat injuries. When the pack is squeezed, an endothermic chemical reaction is activated (see Chapter 9, Chemical reactions, for more on endothermic reactions). This chemical reaction takes in energy from its surroundings and stores this as chemical energy. This leaves its surroundings with less heat energy and therefore the icepack becomes cold

▲ In this bonfire, the chemicals in the wood store energy. When the wood is burned, chemical reactions cause the energy to be released as heat energy

One of the simplest ways of storing energy is to move an object upwards. This could be done by lifting it or moving it up a hill. When the object is raised, work is done against the force of gravity. This requires energy to be transferred and this energy is stored in a form called **gravitational potential energy**. The energy can be released by allowing the object to fall again.

ABC

Gravitational potential energy is the form of energy stored by lifting an object upwards against gravity. It is released when the object falls.

Elastic potential energy is the form of energy stored by stretching or compressing a spring or elastic band.

Kinetic energy is the form of energy possessed by moving objects.

▲ The weights in this clock tower are a store of gravitational potential energy. As they fall, their energy is transferred into maintaining the motion (kinetic energy) of the clock's mechanism, which is losing energy through friction

Stretching or compressing springs or rubber bands requires energy. When a spring is stretched, energy is stored as **elastic potential energy**. When the spring is allowed to return to its original length, the energy is released and can be transferred to kinetic energy or be used to do work.

When objects move, the type of energy they have is called **kinetic energy**. When an unbalanced force acts on an object, it accelerates or decelerates (see Chapter 3, Unbalanced forces). This means that its kinetic energy is changing because work is being done on the object.

▲ The clockwork mechanism of this watch stores elastic potential energy in a spring. As the spring uncoils, the energy is transferred to doing work in moving the cogs and the hands of the watch

▲ This small engine drives two large metal wheels, known as flywheels. These wheels are heavy and store a lot of kinetic energy when they rotate quickly. The purpose of flywheels is to store energy so that when the engine has to do work, the flywheels are able to transfer some energy back to the engine and keep the engine going

ABC **Heat energy** is the form of energy possessed by hot objects.

The temperature of an object is determined by the motion of its particles or molecules (see Chapter 8, States of matter). The more **heat energy** an object has, the faster its particles move, and the higher its temperature. Heat energy can be released in chemical reactions. When there is friction, the work done against friction transfers kinetic energy to heat energy. Heat energy can also be transferred into useful mechanical work with a turbine.

▶ The blade of this angle grinder rubs against the metal. The friction between the two surfaces causes kinetic energy to be transferred to heat energy and the small pieces that are rubbed off the metal by the grinder glow hot as a result

How can energy be transferred?

We have seen that objects may possess or store energy in many different forms. Energy can also be transferred between objects and from one form to another. This can be done through forces: a falling object is converting gravitational potential energy into kinetic energy because of the force of gravity. Likewise, friction can convert kinetic energy into heat energy when two objects rub against each other. There are some other ways in which energy can be transferred.

An electrical current can transfer energy. **Electricity** is one of the most convenient ways of transferring energy because wires can transfer the energy over long distances and to precise electrical components. These components can then transform the energy into the desired form. This is why we use electricity to light our homes and to power many different electrical devices.

Light is a form of **electromagnetic radiation** (see Chapter 6, Heat and light). Light can transfer energy in the form of a wave that has a wavelength that our eyes can detect. Thermal (or infrared) radiation is a similar wave with a longer wavelength. When we feel the heat from the Sun, we are feeling the energy transferred by this thermal radiation. Other wavelengths are responsible for other types of radiation, such as radio waves, microwaves, ultraviolet light, X-rays and gamma rays. All of these can transfer energy too.

ABC **Electricity** is a way of transferring energy using an electrical current.

Electromagnetic radiation is a way in which energy may be transferred in the form of light, heat or other waves.

This laser cutter uses light to transfer energy to the material that it is cutting. Because the laser beam is focused to a small point, a large amount of energy can be directed to a very small area in a controlled way

Energy can also be transferred in the form of sound (see Chapter 5, Waves and sound). The kinetic energy of the molecules in the air is transferred in the form of a wave.

ABC **Sound energy** is a way in which energy may be transferred in the form of sound.

Identifying energy transfers

1. Identify the energy transfers that occur in the following devices.

▲ Loudspeaker

▲ Windmill

▲ Lightbulb

2. For each device, identify the device that works by carrying out the opposite energy transformation.

   For example, a leaf uses energy that is transferred to it via light and stores the energy as chemical potential energy. The reverse process would be for chemical potential energy to be converted into light. A glow stick can do this, or a battery powering a light.

▲ A leaf converts the energy that is transferred to it by light and stores the energy as chemical potential energy

▲ The reverse energy transfer

What happens to their energy when objects collide?

Things can collide in many ways. Sometimes, they bounce off each other. In other collisions the objects may be distorted and deformed, which uses energy. In all of these collisions, energy can be transferred between the objects, but the total amount of energy remains the same.

A Newton's cradle has metal weights that collide. When the weights collide, they transfer energy between them. When the left-hand weight collides, the kinetic energy will be transferred to the right-hand weight. If no (or little) energy is lost, the right-hand weight will gain the same amount of kinetic energy as the left-hand weight loses. This means it will swing up to the same height that the left hand weight started at

In some collisions, no work is done in deforming the objects and no energy is lost as heat. As a result, the objects rebound with the same amount of kinetic energy as they started with (although they may have distributed the kinetic energy between them). These types of collision are called **elastic**.

Other types of collision result in the colliding objects being deformed or gaining heat energy. This results in them rebounding with less kinetic energy than they had before—indeed, they may not rebound at all. These types of collision are called **inelastic**.

ABC

An **elastic collision** is where the colliding objects rebound so that no kinetic energy is lost.

An **inelastic collision** is where the colliding objects rebound at a slower relative speed and so kinetic energy is lost in the collision.

Before:

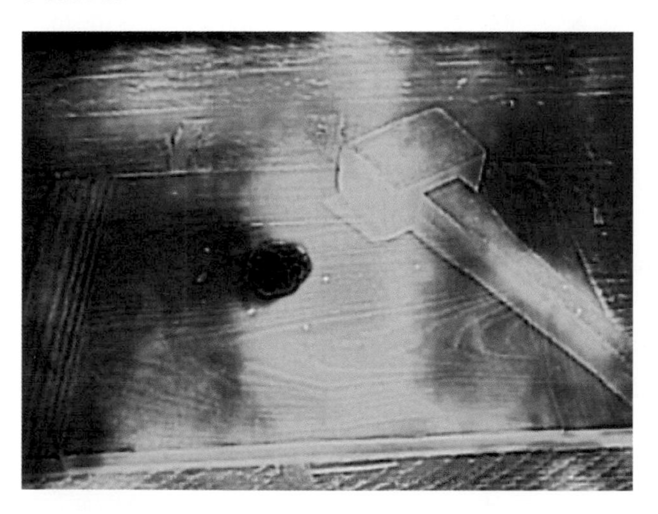

After:

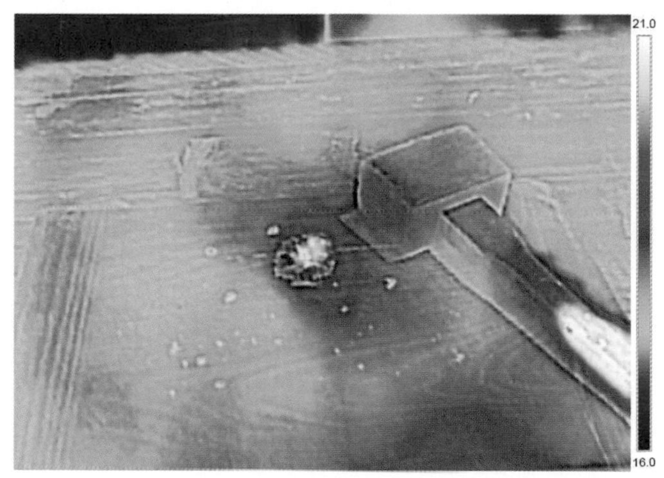

A piece of plasticine was hit with a mallet 100 times. These photographs are taken in infrared light (see Chapter 6, Heat and light) and show the different temperatures. The plasticine started at a temperature of 16°C and ended up at 21°C. This is because each collision did work on the plasticine and transferred the kinetic energy of the mallet into heat energy

Why do things fall?

▲ In a car crash, the kinetic energy of the car is used to do work in deforming the car. The car is carefully designed to do this and absorb as much of the kinetic energy as possible so that the passengers are protected from sudden energy transfers

Falling objects are a good example of energy transfers. Just before the object is dropped, at its highest point, it has gravitational potential energy. As it falls, the gravitational potential energy is converted to kinetic energy. If no energy is lost, then as the object hits the ground, all of the gravitational potential energy will have been converted to kinetic energy. When the object hits the ground, it might retain some of its kinetic energy and bounce, or it might lose all of its energy in deforming the ground or transferring its energy to heat.

In reality, air resistance will oppose the falling object. Work will be done against this force and this will transfer energy away from the falling object. The energy will be transferred to heat energy of the air.

◀ As skydivers fall, they experience air resistance. The work done against the air resistance causes them to lose kinetic energy. At the same time, gravitational potential energy is being converted into kinetic energy. If the gain in kinetic energy from gravitational potential energy happens at the same rate as the loss of kinetic energy from air resistance, then the skydiver will fall at a constant speed. When the skydivers deploy their parachutes, the increased area of the parachute enables them to transfer kinetic energy by doing work on the air more effectively. As a result, they will fall at a slower speed

Diving boards can be at different heights. This pool has diving boards at 10 m, 7.5 m, 5 m, 3 m and 1 m. If a diver dives from the top board, they will start with more gravitational energy. This will be converted into kinetic energy as they fall and since not much energy will be lost to air resistance over the short fall, the diver will hit the water faster as a result

What is gravity?

Because of its large mass, the Earth exerts a force on all objects. In fact, all objects also exert a force between them as well. However, this force between objects is unmeasurably small because most things have a mass that is much smaller than that of the Earth.

The force that the Earth exerts on us is called our weight. The weight of an object can be found using the equation weight = mass × g, where g is the **gravitational field strength**. On Earth, the values of g is 9.8 N/kg.

ABC **Gravitational field strength** is the force exerted on a 1 kg object due to gravity.

If you were to travel to another planet or to the Moon, you would find that the gravitational field strength would be different. This is because other planets have different masses and sizes, and therefore the force with which gravity attracts you is different.

The Apollo 14 mission was the last manned mission to the Moon. This picture, taken 5 February 1971, shows astronaut Alan Shepard hitting a golf ball on the Moon. At the time he claimed it went "miles and miles". If he had been able to hit it properly, then the lower gravitational field and lack of air resistance would have allowed it to travel a long way. However, the limited movement in a space suit meant that he later confessed that it only went a couple of hundred meters

1. Calculate your weight on the Earth.
2. On Mars, the gravitational field strength is only 3.7 N/kg. Calculate your weight on Mars.

What is the conservation of energy?

One of the most important laws of physics is the law of the conservation of energy. While energy may be transferred from one object to another and from one form to a different form, the total amount of energy will always remain the same.

Imagine that a brick were lifted up so that it had 50 J of gravitational energy and then dropped. Halfway down, the brick would have lost half of its original gravitational potential energy. Therefore it would have 25 J of gravitational potential energy remaining. The other 25 J would have been transferred into kinetic energy. As the brick reaches the end of its drop, all 50 J of energy would have been transferred into kinetic energy. As the brick hits the floor, the 50 J of energy is converted into heat energy.

Experiment

How much energy does a bouncy ball lose?

In this experiment you will need a bouncy ball. It could be a basketball, or a tennis ball, or a small toy ball that bounces well.

Method

- Hold the ball 1 meter above a hard floor. Measure this distance from the floor to the bottom of the ball.
- Drop the ball and measure how high the ball bounces. Again, you should measure this height to the bottom of the ball.
- Repeat this measurement three times.
- Take these measurements for a range of heights. 0.1 m up to 1 m in steps of 0.1 m might be sensible, but it will depend on how well your ball bounces. Aim for at least 7 different heights that are easily measurable.
- Record your data in a table and plot a graph of your results.

Questions

1. The fraction of energy lost in each bounce is equal to the fraction $\frac{\text{height of bounce}}{\text{original height}}$. What fraction of energy was lost in each bounce?
2. How does the height at which the ball is dropped affect the fraction of energy lost?

Summative assessment

Statement of inquiry:

Changes in energy drive the basic processes of nature.

This assessment is based around energy considerations in cars and road safety.

Energy, gravity and cars

1. A driver parks his car on a hill but forgets to put the brakes on. The car starts to roll down the hill. Which of the following energy transformations occurs as the car rolls down the hill?

A. kinetic energy → gravitational potential energy

B. kinetic energy → heat energy

C. gravitational potential energy → heat energy

D. gravitational potential energy → kinetic energy

2. When a driver wants to stop a car, he presses the brake pedal. The brakes are applied, which use friction to slow the car to a stop. Which of the following energy transformations occurs as the car brakes?

A. kinetic energy → gravitational potential energy

B. kinetic energy → heat energy

C. gravitational potential energy → heat energy

D. gravitational potential energy → kinetic energy

3. A toy car has a kinetic energy of 10 J. It rolls up a slope and at the top it has 3 J of gravitational potential energy and 5 J of kinetic energy. How much energy has the car lost through friction?

A. 2 J **B.** 3 J **C.** 5 J **D.** 7 J

4. A car travels along a flat road. It has 500 billion J of kinetic energy. Every second, air resistance and other frictional forces cause it to lose 1 billion J of energy. If the car is to travel at a constant speed, how much energy will the engine have to convert to kinetic energy every second?

A. 1 billion J

B. 499 billion J

C. 500 billion J

D. 501 billion J

5. One day, car journeys may be possible on the Moon. Which statement is correct about a car on the Moon?

A. A car would not work on the Moon as there is no gravity.

B. A car traveling on the Moon would have less kinetic energy than a car traveling at the same speed on Earth.

C. A car would need more energy to climb a hill on the Moon than the same car climbing a similar hill on Earth.

D. A car on the Moon would have less weight than an identical car on Earth.

Investigating stopping distances

A pupil designs an experiment to investigate the energy transfers involved with a toy car. In the experiment, they roll a toy car down a ramp and then see how far it travels on a flat surface before frictional forces do enough work to bring it to a stop. The pupil decides to investigate how changing the mass of the car affects the distance it travels after the ramp.

The pupil writes the following method for this experiment.

- Make a ramp and mark a starting line 1 m from the base of the ramp.
- Position the car on the start line so that it is pointing straight down the ramp.
- Release the car and allow it to roll down the ramp and along the flat surface until it comes to rest.
- Measure the distance that the car travels from the base of the ramp.
- Repeat two more times and take an average of the three distances.
- Add some toy putty to the car to make it heavier. Repeat the experiment with increasing amounts of putty. Each time, take three measurements of the distance the car travels.

6. Identify the independent and dependent variables in this experiment. [2]

7. Suggest one control variable for this experiment. [1]

8. The pupil carries out the experiment and realizes that there is one measurement that he needs to take that is not mentioned in his method. Which measurement is this? [1]

9. Which pieces of apparatus will the pupil need in order to take measurements? [2]

10. Write a suitable hypothesis for this investigation. Explain your hypothesis with scientific reasoning. [3]

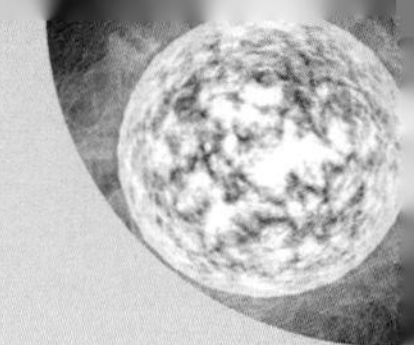

Speed management

In 2008, the Global Road Safety Partnership published a report titled *Speed Management*, which investigated the effects of speed in road crashes. In it they referenced a report from the Organisation for Economic Co-operation and Development, also titled *Speed Management*, which was published in 2006. This report referenced other research on the probability that an accident was fatal.

11. The reports reference other sources. What information is required in a reference that would enable you to find the original source of information? [3]

The report investigated the case of a pedestrian being hit by a car. In this situation they examined how the speed of the car affected the probability of the pedestrian surviving the crash.

12. Explain, in terms of energy, why a car traveling at a faster speed is more dangerous in a collision. [1]

13. Identify the independent and dependent variables in this investigation. [2]

14. Explain why the researchers would not carry out experimental research in this investigation. [2]

The conclusions of the research are shown below.

Speed (km/h)	Probability that crash is fatal (%)
10	4
20	5
30	11
40	36
50	85
60	98
70	99

15. Plot a graph of the data. [5]

16. Some states have residential speed limits of 30 mph (48 km/h). Other states have speed limits of 20 or 25 mph (32 or 40 km/h). Use your graph to find the probability that a crash is fatal at these speeds. [3]

17. Use your data to write a short article to persuade people of the benefits of cutting residential speed limits from 30 mph to 25 or 20 mph. [4]

5 Waves and sound

Key concept: Relationships

Related concepts: Models, Interaction

Global context: Personal and cultural expression

There are many different types of wave. Gravitational waves were predicted by Einstein's general theory of relativity, published in 1916. However, they are so weak that they were not detected for almost a century. The Laser Interferometer Gravitational-Wave Observatory (LIGO) in Livingstone, Louisiana, USA is one of two facilities built to detect gravitational waves. They started collecting data in 2002 and, in 2016, they announced the first observation of gravitational waves. Why do physicists build such huge and expensive experiments?

A crowd at a football stadium can perform a Mexican wave. Each person stands and waves their hands in the air when they see the person next to them doing the same. This wave travels around the stadium, even though the spectators remain at their seats. Where else do people or animals participate in coordinated movements?

▶ Elephants use infrasound to communicate across distances of several kilometers. These sounds have frequencies that are too low for our ears to detect. Which other animals use sounds that are outside our range of hearing?

Statement of inquiry:

The waves that we see and hear help to form our relationship with the outside world.

▶ X-rays are waves that are very similar to light except that they have a very short wavelength. This means that they can pass through many materials. We use X-rays to see inside the body. The X-rays are not stopped by soft tissue, but they are blocked by bone. Therefore this X-ray of a snake shows the snake's skeleton. How else do we use waves in medicine?

Key concept: Relationships

Related concepts: Models, Interaction

Global context: Personal and cultural expression

Statement of inquiry:

The waves that we see and hear help to form our relationship with the outside world.

Introduction

Chapter 4, Potential energy, kinetic energy and gravity, shows how energy can be transferred by waves. These waves not only transfer energy, such as the heat and light that we get from the Sun, but they can also transfer information. Indeed, our senses of sight and hearing, which help us to understand the world around us, rely on detecting these waves. As a result, one of the related concepts of the chapter is interaction.

There are many examples of waves. In this chapter, we see how light waves and sound waves behave and how they are like the waves that travel along ropes or strings, or that we see on the surface of water. These are useful comparisons to help us understand other types of wave, and therefore the other related concept is models. Just as we use the light and sound that we see and hear to create a model of the world around us, we can use waves to investigate other things and create models of them.

Our senses rely on waves to understand the world around us, but we also use them to communicate and interact with each other. For this reason, the global context of the chapter is personal and cultural expression. We will see how waves enable us to communicate and how they can be useful to us.

▲ This is a sonar image of a patrol boat that was sunk in World War II. Sonar images use the reflections of sound waves to build an image. Using sonar enables investigators to see an image of the wreck, which is about 100 m under the water's surface, without having to send divers down

What is a wave?

If a rope is laid along the ground and the end is flicked quickly up and down, a wave will travel along the rope. Although parts of the rope will move up and down, returning to where they started, the effect is that you see the disturbance move along the rope.

The waves that can be created on a rope or a string are a good example of a wave. Waves are very common in nature and although many waves are very different to those created on a rope, these waves are a good way of understanding what waves are and what they can do.

A wave can be created on a rope by moving one end up and down. This is a popular exercise at gyms and park fitness sessions

Although waves involve movement, they do not move matter from place to place. Instead, any motion of the material that a wave travels through will result in the matter returning to its original place. In this example, the wave travels down the rope, but the rope itself will return to its original position.

Although waves do not transfer matter, they can transfer energy. If a small insect was crawling down the rope as the wave passed, it would be thrown upwards by the passing wave.

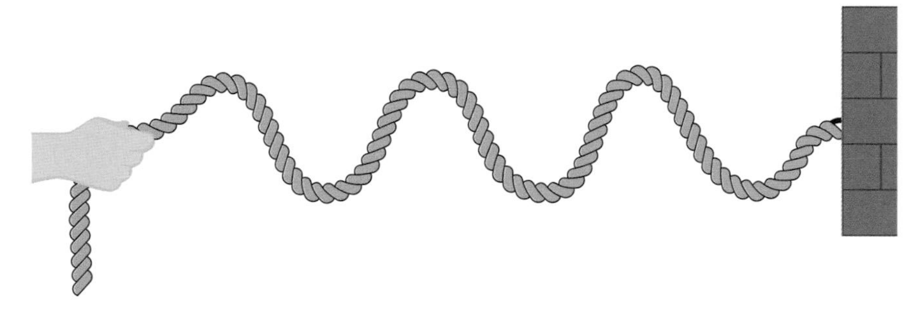

Moving the end of a rope up and down can cause a wave to travel down the rope

How do waves travel on water?

You will already be familiar with waves that travel across the surface of water. Small waves, or ripples, are often seen on the surface of water, such as a pond. These are a good example of how waves work. The level of the water goes up and down, and the disturbance travels along the surface of the water. The overall level of the water is unchanged, and the wave does not cause water to move from one end of the pond to the other.

The waves that wash up on the beach are not such a simple example. As waves approach the shore, the depth of the sea becomes less, and the waves start to slow down. This is why waves break as they reach the shore. Because the water is shallower, the waves cause water to move up and down the beach. This is not how waves behave in deep water—such as in a pond, where the water level moves up and down, but no water is moved along by the wave.

As waves reach the shore, their speed changes, causing them to break. This makes their behavior complicated and therefore these waves are not such a good model of how most waves behave. In the deeper sea, waves travel along transferring energy, but not moving water along with the wave

Waves transfer energy. This picture shows a wave energy generator, which converts the energy of the waves on the ocean into electricity. As the ocean waves pass, they cause the joints between the different sections to bend and this action is used to convert the transferred energy of the wave into electrical energy

What types of wave are there?

There are many other types of wave apart from the waves that travel along the surface of water and down ropes and strings. These other types of wave can travel through many different materials and seem to be very different. However, they all transfer energy and behave as waves.

Sound travels as a wave. The particles in the air move back and forth and allow the wave to travel through the air. Although a sound wave can transfer energy, the air itself does not move along with the wave.

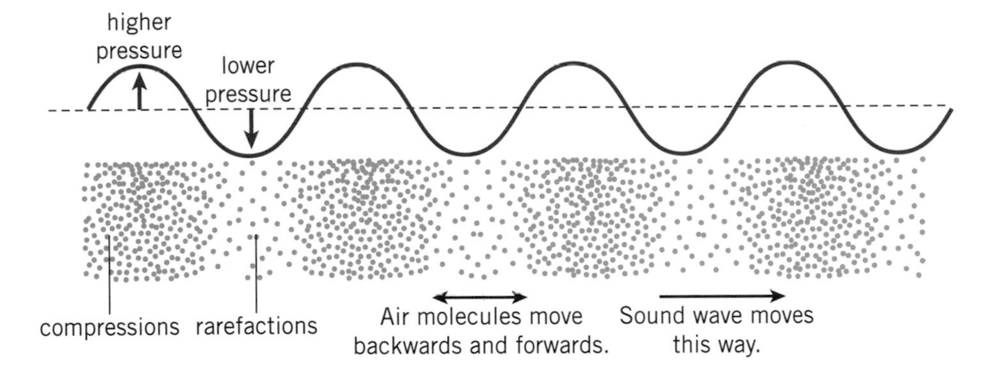

▲ In a sound wave, the air molecules move backwards and forwards, bumping into each other and creating a wave that moves through the air

Earthquakes cause waves to move along the surface of the Earth as well as traveling down towards the center of the Earth. These waves are called **seismic waves**.

The waves involved in sound, earthquakes and on the surface of water are **mechanical waves**. This is where the wave is caused by matter moving up and down or backwards and forwards. These waves cannot travel without the matter (or **medium**) through which they move. This is why sound cannot travel through a vacuum such as space as there are no air particles to allow the wave to travel.

There are some waves, such as light, that do not need a medium to travel through. There are many other waves that behave in the same way as light: radio waves, microwaves, infrared, ultraviolet, X-rays and gamma rays are all waves that can pass through a vacuum.

ABC

Seismic waves are waves that travel through the ground caused by earthquakes.

A **mechanical wave** is a wave caused by the physical movement of something.

A **medium** is the material through which a wave travels.

How do waves behave?

When waves meet a barrier, they may be able to pass through it or they may bounce back. Whether they pass through depends on the type of wave and what the barrier is. For example, light waves can pass through the glass in a window, but sound waves will mostly bounce back. When waves bounce off a surface they are said to reflect.

▶ The Sun's heat and light travel as waves from the Sun to the Earth. This type of wave does not need a medium to travel through and therefore the waves are able to reach the Earth. Any sound that the Sun makes, however, would not be able to travel through space to the Earth

▲ In a recording studio, the window allows the light waves to pass through so that the sound engineer can see what is happening in the studio. The sound waves do not pass through as much and therefore any sound that the engineer makes does not get recorded

ABC A **reflection** is the act of waves bouncing off a surface.

An **echo** is an effect caused by the reflection of sound waves.

Reflection is a property that is often associated with light. However, many other waves can reflect. When sound waves reflect, they cause an **echo**. In Chapter 6, Heat and light, we look at how waves of light are reflected.

▶ Seismic waves are caused by earthquakes. Some waves travel along the surface of the Earth, others travel into the center of the Earth and reflect off the iron core. While a large earthquake can cause large amplitude waves, seismic waves with small amplitudes are very common. A device called a seismometer is used to detect seismic waves. Careful measurement of these seismic waves can enable scientists to determine where an earthquake started (called the epicenter). Study of seismic waves can also be used to monitor volcanoes and even deduce the structure of the inside of the Earth by measuring the echoes of the waves off the iron core (see Chapter 10, The Earth)

1. S-waves are a type of seismic wave that are created by earthquakes. These waves travel at about 3 km/s. How far would they travel in 1 hour?

Experiment

Measuring the speed of sound waves

The speed of sound varies depending on the weather, altitude and temperature. However, it is usually around 300 m/s. To measure the speed of sound, you need a stopwatch and a way of making a loud sound. A clapperboard is a good way of doing this, as it uses two pieces of wood that are connected with a hinge and can be slammed shut.

Method

1. Find a large surface for the sound to reflect off. The side of a building is ideal for this. Ideally you should be about 100 m away from the building and there shouldn't be other buildings nearby for the sound to reflect off (otherwise you will hear many echoes).
2. Measure the distance from you to the building.
3. Make a loud sound with the clapper board.
4. Working in pairs or a small group, have one person make the sound, while the others measure the time interval between making the sound and hearing the echo.

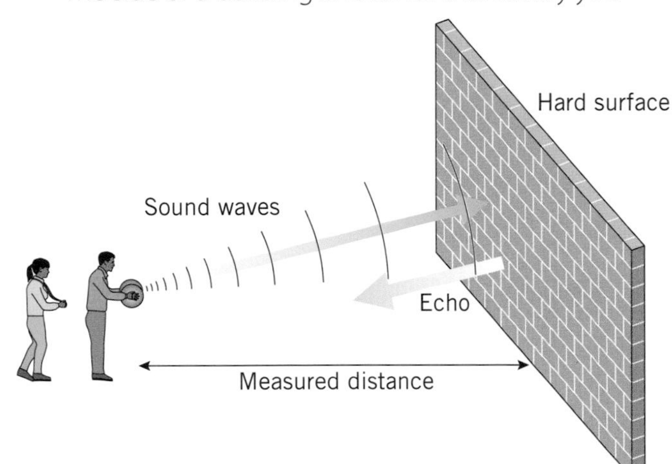

Questions

1. Calculate the speed of the sound using the equation:

$$\text{speed} = \frac{\text{distance}}{\text{time}}$$

Don't forget that the distance in your calculation should be the distance to the wall and back again.

2. How reliable is your measurement? How could you improve your measurement in order to get a more accurate answer?

What happens when two waves meet?

When two waves meet, they pass through each other. At the point where they meet, they will add together. The addition of waves is called **interference**. Waves can add together in different ways. If both waves are acting in the same direction, i.e. if the crest of one wave overlaps with the crest of the other wave, then the result will be a larger wave than before. We call this **constructive interference**. Alternatively, if the waves are acting in opposite directions such that the crest of one wave overlaps with the trough of another, then they will cancel each other out. The result is a smaller wave, or no wave at all. This is called **destructive interference**.

ABC

Interference is the effect of waves adding together.

Constructive interference is where two waves add together to form a larger wave.

Destructive interference is where two waves cancel each other to give a smaller resulting wave.

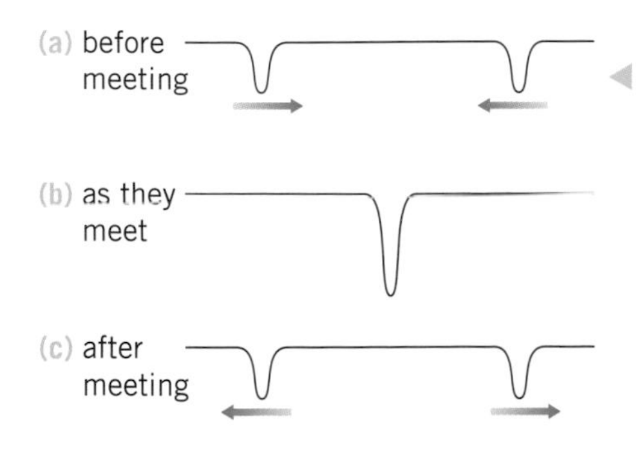

(a) Two waves travel along a rope or along the surface of water towards each other

(b) Where they meet, they add together in an effect called interference. Here, the two waves add together to make a larger wave. If the original waves had been in opposite directions, they would have canceled each other to give a smaller wave, or no wave at all

(c) The two waves only overlap for an instant. They move through each other and continue in their original directions

Demonstrating how waves add together

This is quite tricky to get to work, but an effective demonstration when it does work. Lay a slinky on a long table. Balance some pens on their ends to make two lines on each side of the slinky to act as markers. Felt tip pens, clothespins or small lengths of wood can be used.

- Send a single wave from either end of the slinky so that both pulses act in the same direction. Each wave should be small enough that it does not knock over the markers.

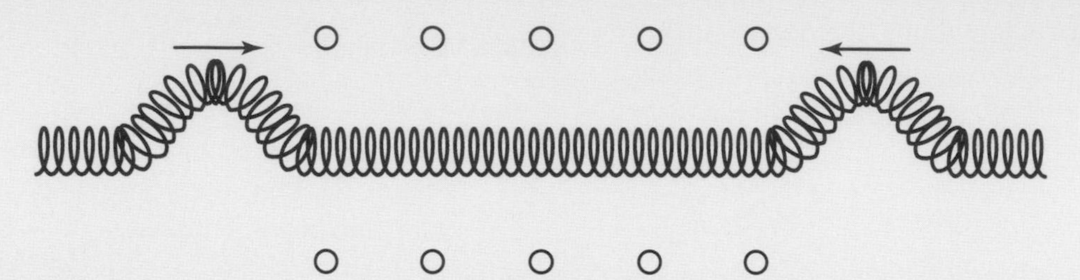

Where the two waves meet, a larger wave is created that can knock over the marker.

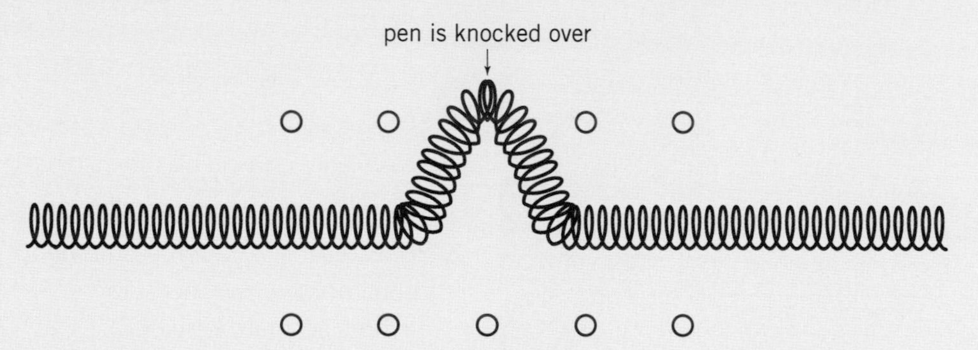

- Now move the markers closer to the slinky and send two pulses in opposite directions from the ends of the slinky. This time the waves should be large enough to knock over the markers.

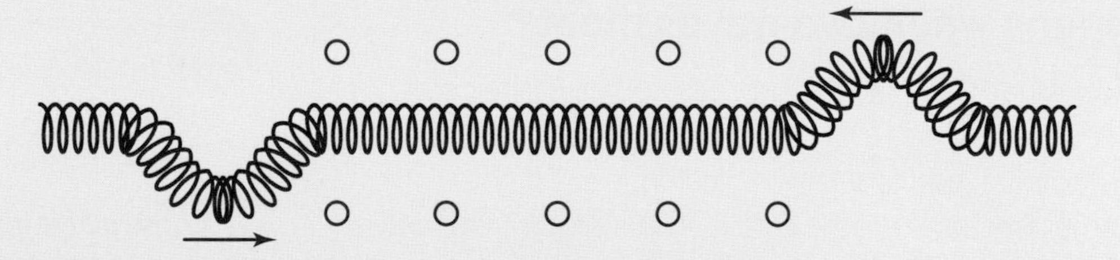

The markers will be knocked over, but where the two waves add together and cancel each other out, the markers can remain upright.

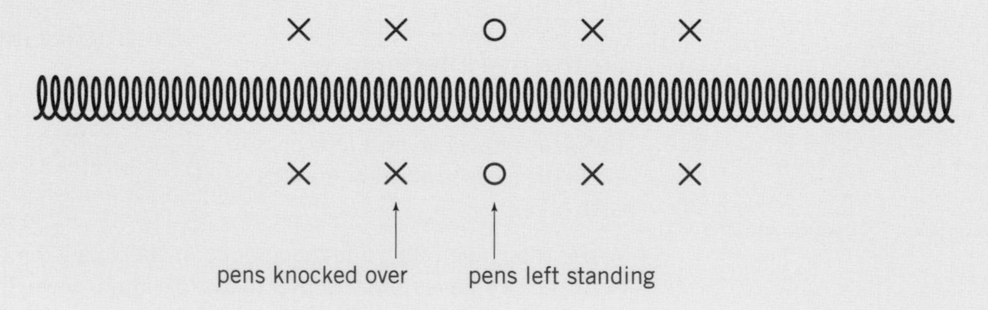

How can we describe waves?

Waves fall into one of two categories: longitudinal and transverse.

Imagine a wave traveling across the surface of a pond. Although the wave moves across the surface, the water itself moves up and down at right angles to the surface. As a result, something floating on the surface of the water would bob up and down as the wave passes under it. This type of wave is called a **transverse wave**. These up and down motions (or oscillations) are at right angles to the direction in which the wave travels. Light is a good example of a transverse wave. Radio waves, microwaves, infrared, ultraviolet, X-rays and gamma rays are also transverse waves. These particular examples are part of something called the **electromagnetic spectrum**.

In a **longitudinal wave,** the oscillations are parallel to the direction in which the wave travels. Sound waves are good examples of longitudinal waves. The particles in the air move back and forth, colliding with each other and as they do so, a sound wave is able to travel through the air.

Longitudinal and transverse waves can be demonstrated using a slinky spring. If you hold one end of a stretched slinky and shake your hand from side to side, you will create a transverse wave. Moving your hand forwards and backwards along the line of the slinky will create a longitudinal wave.

ABC

Transverse waves have oscillations at right angles to the direction in which the wave is traveling.

The **electromagnetic spectrum** is a group of transverse waves that includes radio waves, microwaves, infrared, visible light, ultraviolet, X-rays and gamma rays.

Longitudinal waves have oscillations parallel to the direction in which the wave is traveling.

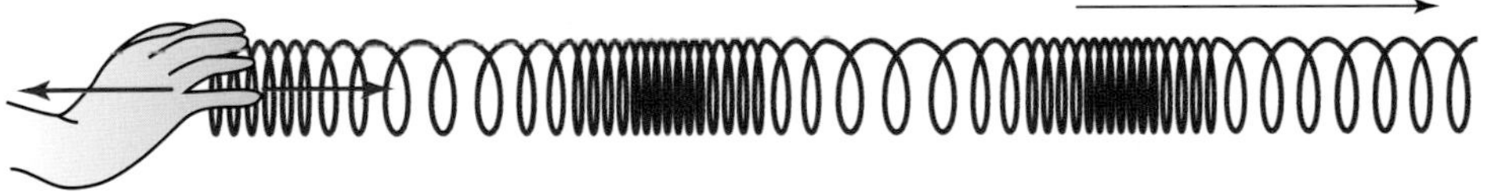

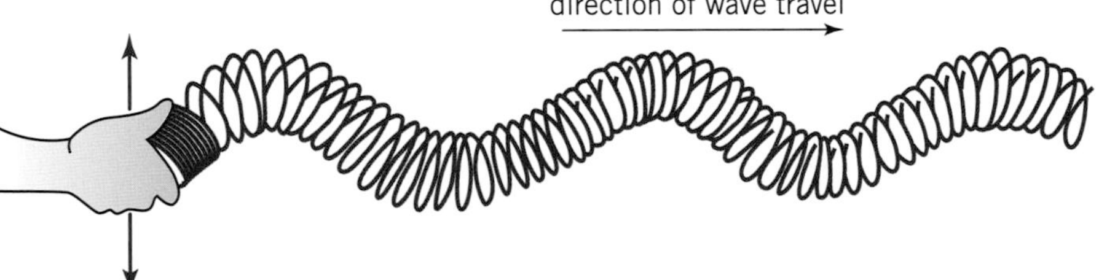

A longitudinal wave is made by moving your hand forwards and backwards. The coils in the slinky move parallel to the direction in which the wave travels. Moving your hand side to side causes a transverse wave because the slinky's coils move at right angles to the direction in which the wave travels

As a wave passes under the duck and its ducklings, they are moved up and down. This is because the waves along the surface of the water are transverse waves

ABC The **amplitude** is the maximum displacement of a wave.

The **frequency** is the number of waves that pass in one second.

The **time period** is the amount of time a complete wave takes to pass.

How can we measure waves?

Waves of the same type can still differ. There is a big difference between small ripples on the surface of a pond and large waves in the ocean, even though these are both waves on the surface of water. It is important to be able to describe these differences.

One way in which waves can differ is in their **amplitude**. The amplitude is a measure of how far a wave rises or falls from its starting point. As a result, a high amplitude wave will carry more energy than a low amplitude wave.

Another important property of a wave is its **frequency**. Frequency is a measure of how many full oscillations occur in one second or how many waves pass a point in one second. The units of frequency are hertz (Hz).

The frequency of a wave is also related to the **time period** of a wave. The time period is how long it takes each wave to pass a point. The frequency and time period are related by the equation:

$$\text{frequency} = \frac{1}{\text{time period}}$$

For example, a wave has a time period of 0.2 s. To calculate its frequency, you can use the above equation.

$$\text{frequency} = \frac{1}{\text{time period}} = \frac{1}{0.2} = 5\text{ Hz}$$

ABC The **wavelength** is the distance from one wave to the same point on the next wave along.

The **wavelength** of a wave is a measure of the distance from the peak of one wave to the peak of the next. The wavelength and frequency are related: if a wave has a short wavelength, the frequency will be higher, and if a wave has a longer wavelength then the frequency will be lower.

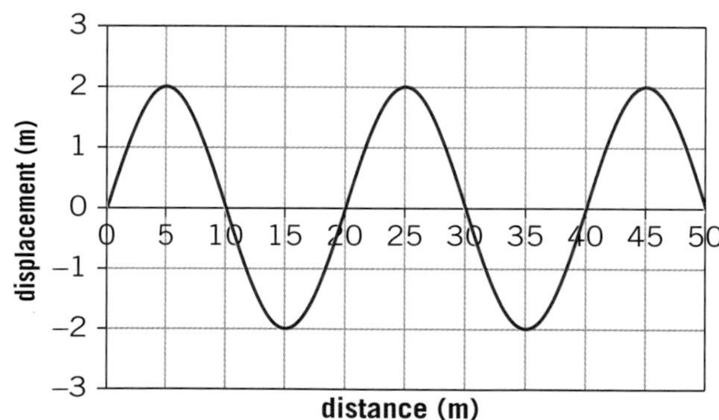

1. The graph on the right shows the displacement of water on a lake. Use the graph to find:
   a) the amplitude of the waves
   b) the wavelength of the waves.
2. The lowest note on a piano has a time period of 0.036 s. Calculate the frequency of this note.

How can we measure sound?

Sound waves are longitudinal. Like all waves, they have a wavelength, a frequency, a time period and an amplitude. Different wavelengths, frequencies and amplitudes will mean that sound waves sound different.

A sound wave with a large amplitude carries more energy and sounds louder. A quiet sound has a small amplitude. Sounds with a high frequency and short wavelength will sound higher in pitch while low frequency sounds are lower in pitch.

Because smaller objects are able to vibrate faster than larger objects, they are able to make higher pitched sounds. This is why large musical instruments tend to make low pitched sounds while smaller instruments make higher pitched sounds.

How do musical instruments change their note?

We have seen that smaller objects are generally able to make sounds with a smaller wavelength and therefore have a higher pitch. Look at the pictures of these instruments. Identify what the player of each instrument would have to do to change note. How does this affect the size of whatever is making the note?

Tuning forks

Tuning forks are used to create a note that other instruments can tune to. They come in different sizes to create different notes. A pupil finds the set of tuning forks shown in the photograph. They produce the notes from middle C to upper C. Each tuning fork is marked with the note it makes, as well as its frequency.

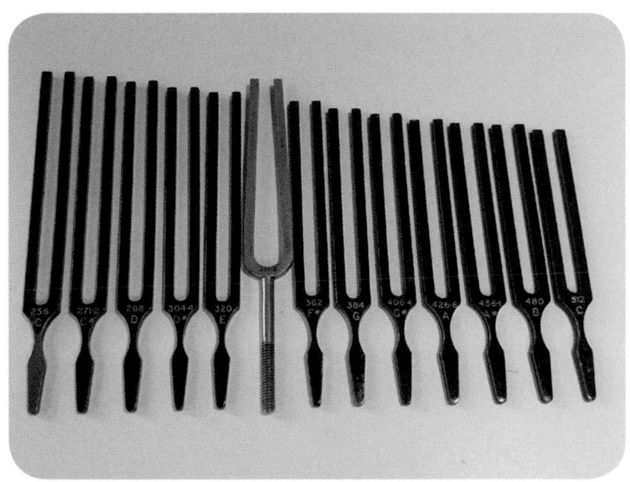

The pupil measures the mass of some of the tuning forks and creates a table, shown on the right.

Note	Frequency (Hz)	Mass (g)
C	256	47.44
D	288	46.30
E	320	43.76
F	341.3	61.44
G	384	40.58
A	426.6	38.41
B	480	37.03
C	512	36.73

1. Using the table, plot a graph of frequency against mass for the set of tuning forks.
2. One of the points in the table is an anomaly. Suggest which point and give a reason for the anomaly.
3. Using your graph, describe the relationship between the frequency and the mass of the tuning forks.
4. The G# tuning fork has a frequency of 406.4 Hz. Using your graph, predict the mass of the G# tuning fork.

The longer strings on a piano vibrate at a lower frequency because they are longer and thicker. This means that they make a lower sound than the shorter strings, which make high pitched sounds. Different lengths of string create the different notes

Humans can hear sounds between the frequencies of 20 Hz and 20,000 Hz. Some sounds have higher frequencies than we can hear. We call these sounds ultrasound. Although humans cannot hear these sounds, some animals such as bats can use ultrasound to navigate in the dark. This is called echolocation. Ultrasound can also be used for medical imaging.

Other animals are able to hear sounds that are lower than 20 Hz. These sounds are called infrasound. Elephants can communicate over long distances by using infrasound and some birds, such as pigeons, have been shown to detect infrasound. Studying low frequency sounds can also be used to monitor volcanoes. Indeed, some studies have suggested that the infrasound emitted by natural occurrences such as storms, earthquakes and volcanoes can be heard by animals as a warning sign.

Sounds that are above the range of human hearing are called ultrasound. These sounds have wavelengths that are too short and frequencies that are too high for our ears to detect. Some animals such as bats use ultrasound to communicate and they use the reflections from the ultrasound to navigate in the dark and find insects to eat

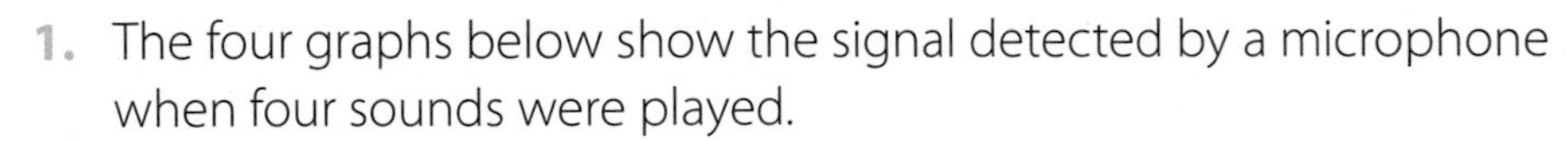

1. The four graphs below show the signal detected by a microphone when four sounds were played.

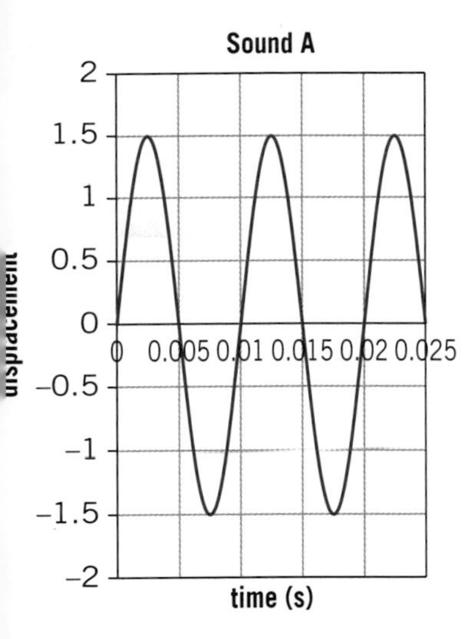

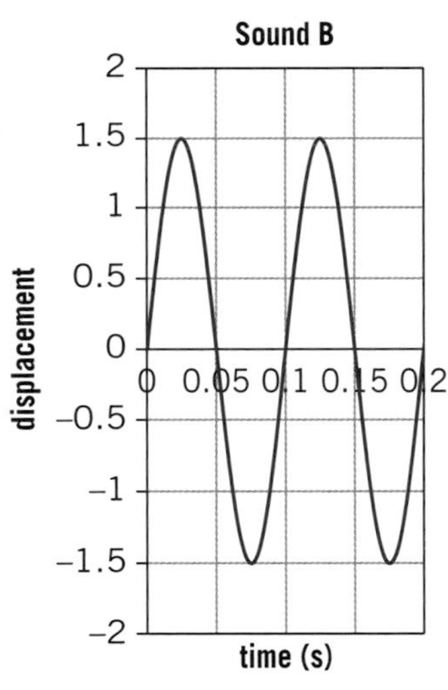

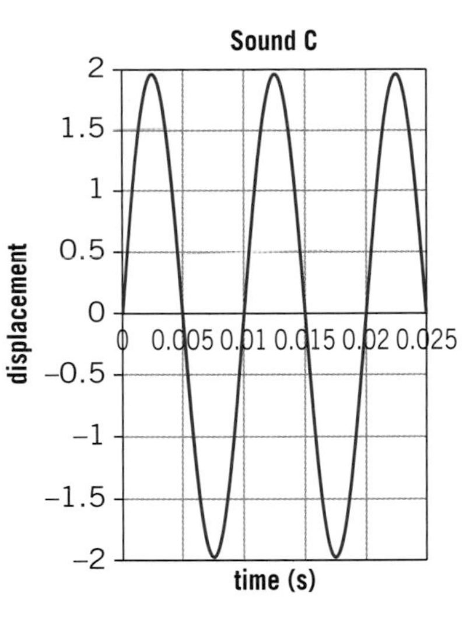

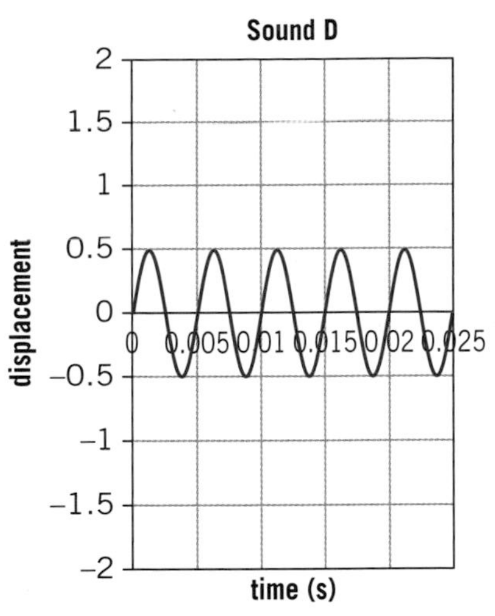

Identify which graph represents the sound that

a) is the loudest

b) is the highest in pitch

c) cannot be heard by humans.

How do different wavelengths and frequencies affect light?

Changing the frequency of a sound wave will make it sound at a different pitch and changing the amplitude will change the volume. Changing the frequency and amplitude of a light wave will also affect how the light is seen. Increasing the amplitude of a light wave will increase the amount of energy that is carried and the light will appear brighter.

Changing the wavelength and frequency of light will change its color. Red light has a longer wavelength and lower frequency than blue light. In between red and blue light, our eyes detect the different colors that are seen in a rainbow. From the longest wavelengths to the shortest they are red, orange, yellow, green, blue, indigo and violet.

Light waves can have wavelengths and frequencies that fall outside of the range that our eyes can detect. Light with a higher frequency and shorter wavelength is called ultraviolet radiation, and waves with longer wavelengths and lower frequencies are called infrared radiation.

The colors of visible light

The table below shows the colors of the visible spectrum with the typical wavelengths of that color of light measured in nanometers (1 nanometer is a billionth of a meter or 10^{-9} m).

Color	Wavelength (nm)
red	625–740
orange	590–625
yellow	565–590
green	520–565
blue	445–520
indigo	425–445
violet	380–425

1. Construct a number line, like the example below, from 300 to 800 nm and label the colors. Green has already been labelled for you.

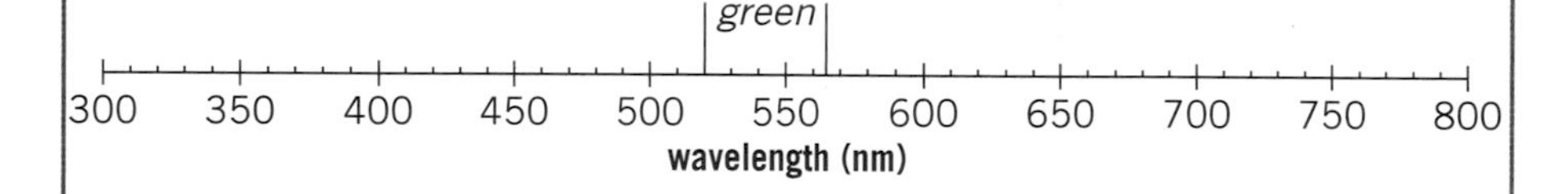

2. Color in the regions to produce the colors of the rainbow.
3. Mark on your diagram where infrared and ultraviolet radiation occur.
4. Your eyes have special cells that detect different colors of light. These cells are called cones. These cones come in three types that are sensitive to different wavelengths. The wavelengths of light that each type of cone is most sensitive to are 420 nm, 534 nm, and 564 nm. Which colors are these?

Sometimes, light is split into a spectrum or rainbow, which shows the different colors of light with different wavelengths

ATL **Communication skills**

Understanding human perception

We rely on waves to perceive the world around us. We feel infrared waves from the Sun as heat, we hear sound waves and we see visible light. As much as we rely on the information that the waves transmit to us, our brains are responsible for converting these signals into information that we can understand. This final stage in our perception can be unreliable: our brains try to make sense of what we see and hear but in interpreting this information, we may perceive something that isn't really there.

Optical illusions are a good example of our sight being fooled as our brains interpret the information. There are many other examples of our senses being fooled. For example, it has been shown that the perceived taste of food can depend on its color, or even the color of the plate it is served on.

By understanding how the brain can be fooled, scientists and psychologists can manipulate our senses. Adverts can be made visually appealing and people can be trained to speak in ways that makes them sound trustworthy or motivational. This knowledge is not just used for deception—it can also be used to inspire feelings of wellbeing and positive moods. Can you think of any examples of this?

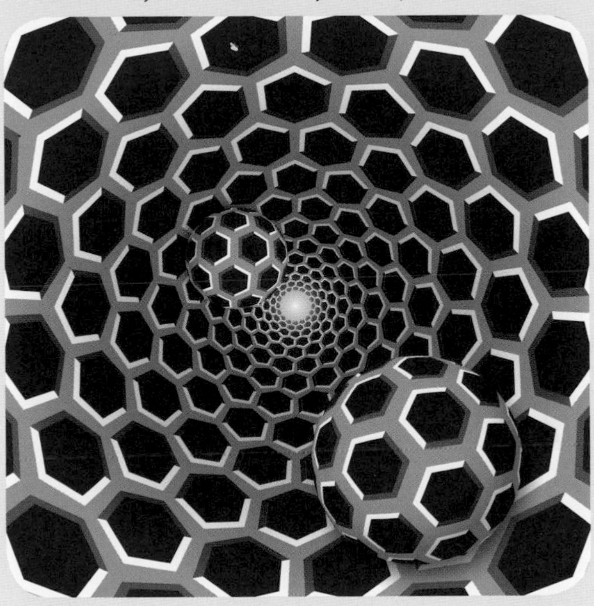

Look at these pictures. They trick your brain into thinking that they are moving

Summative assessment

Statement of inquiry:

The waves that we see and hear help to form our relationship with the outside world.

Water covers about 71% of the Earth's surface. The vast majority of this water is in the seas and oceans. Wind blows across the surface of seas oceans and lakes and causes waves to form. Waves can also be caused by earthquakes, which can result in dangerous tsunamis. This assessment is based on waves that travel across the surface of the ocean.

Waves and sound

1. A low note has a time period of 0.0125 s. What is its frequency?

A. 8 Hz

B. 12.5 Hz

C. 80 Hz

D. 125 Hz

2. Which of the following is not a transverse wave?

A. The ripples on the surface of the ocean

B. The ultrasound waves that dolphins emit to find fish

C. The infrared waves given off by the warm surface of the ocean

D. The visible light that reflects off the surface of the ocean

3. Here are four statements about the difference between the light waves we see and the sound waves we hear. Which of the following statements is true?

A. Sound waves are transverse but light waves are longitudinal.

B. Sound waves can travel through a vacuum but light waves cannot.

C. Sound waves have a much lower frequency than light waves.

D. Sound waves cannot transfer energy but light waves can.

4. Two identical stones are dropped into opposite ends of a puddle. The waves created by the stones travel towards each other. What happens in the middle of the puddle?

A. The waves reflect off each other.

B. The waves pass through each other.

C. The waves add together to form one larger wave, which travels to the end of the puddle.

D. The waves cancel each other out and the result is that there are no waves.

5. The following graphs show the signals recorded by a microphone for two different sounds, A and B.

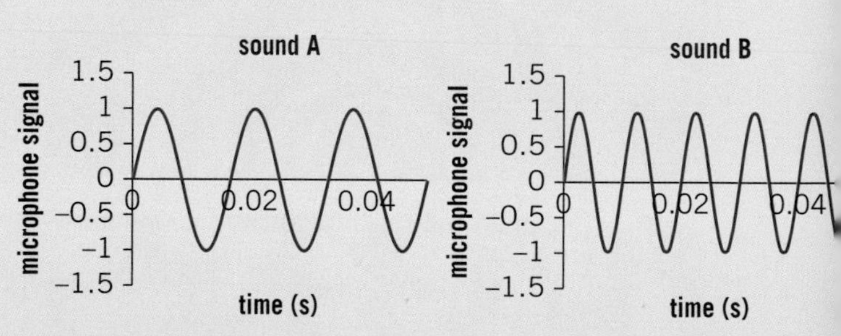

Which statement is true?

A. Sound A is louder than sound B.

B. Sound A is quieter than sound B.

C. Sound A is higher pitched than sound B.

D. Sound A is lower pitched than sound B.

6. Which property cannot be transferred by a wave?

A. Energy

B. Information

C. Heat

D. Matter

7. Which of the following does not involve a wave bouncing off a surface?

A. An echo

B. A full moon

C. A shadow

D. A reflection in a mirror

Does salt water cause waves to travel faster?

A student is curious to know whether waves travel at different speeds in salty water or fresh water. He thinks that saltier water will have a greater density (due to the salt dissolved in it) and that the force of gravity that pulls the water back to its normal level will be greater. As a result, he makes a hypothesis that waves will travel faster in saltier water.

To carry out his experiment, he gets a large, circular paddling pool. He writes down the following method:

- Fill the paddling pool to a depth of 20 cm with fresh water.
- Measure the distance from the center of the pool to the outside rim.
- Drop a stone into the center of the paddling pool and time how long it takes for the wave to reach the edge of the paddling pool.
- Repeat this measurement three times.
- Add 2 kg of salt to the paddling pool and stir until it is all dissolved.
- Drop the stone into the center of the paddling pool and time how long it takes for the wave to reach the edge of the paddling pool. Repeat this three times as before.

8. Identify the independent and dependent variables in this experiment. [2]

9. Identify one control variable in this experiment. [1]

The results of the experiment are shown in the table below.

Type of water	Time 1 (s)	Time 2 (s)	Time 3 (s)
fresh	5.12	5.15	5.30
salty	5.09	5.12	5.34

10. The distance to the edge of the paddling pool was 1 m. Calculate the average speed of the waves in fresh water and salt water. [2]

11. Discuss whether the results support the student's hypothesis. [3]

12. Suggest and explain an improvement to the experiment. [2]

The speed of waves on water

A student researches the speed that waves travel on water. He plots a graph of the speed against the wavelength.

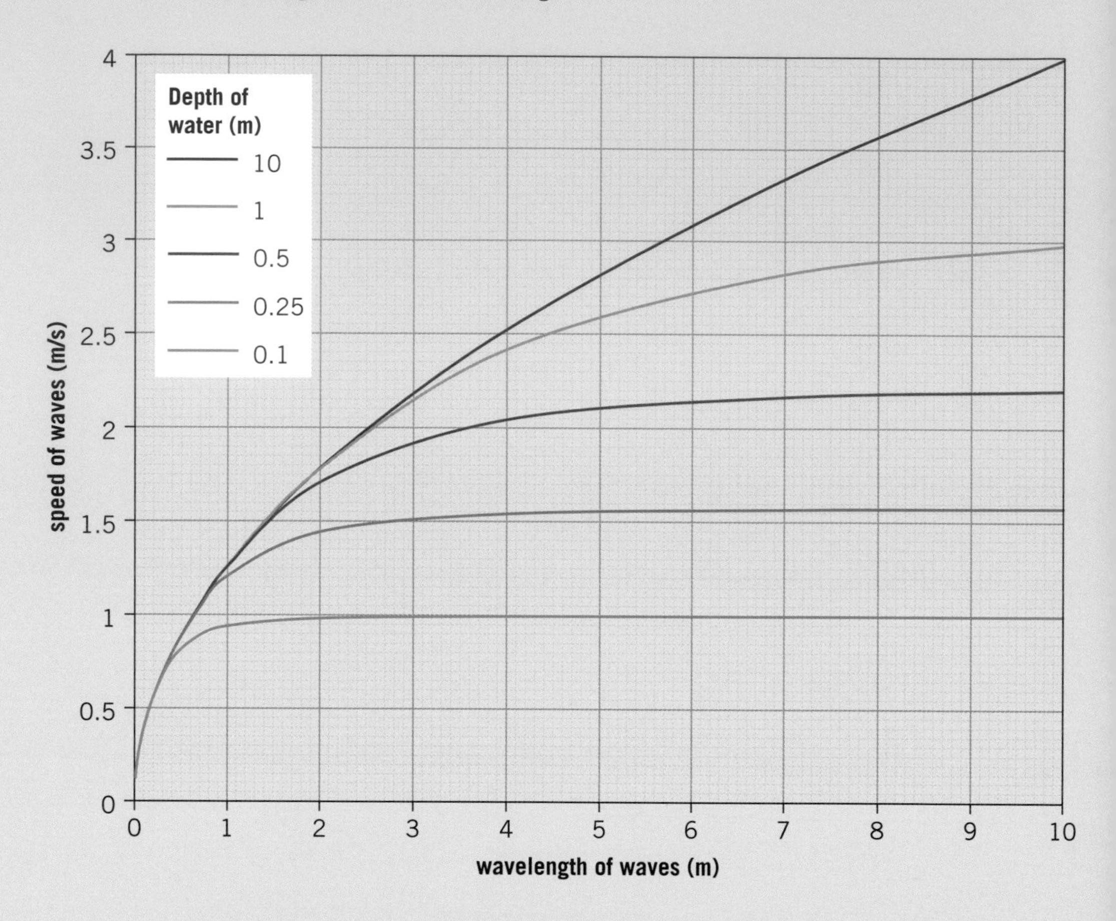

13. Use the graph to estimate the speed of waves that have the following wavelengths in these depths of water:

a) a wavelength of 0.5 m in a depth of 5 m [1]

b) a wavelength of 20 m in a depth of 0.1 m [1]

c) a wavelength of 2 m in a depth of 5 m. [1]

The student draws the following conclusions.

- Waves always travel faster in deeper water.
- Increasing the wavelength of the wave would make it travel faster.
- As long as the wavelength is at least 20 times longer than the depth of the water, the wavelength does not affect the speed at which the wave travels.
- If the wavelength is smaller than the depth of the water, then the depth of the water does not affect the speed of the waves.
- As waves from the ocean reach the shore, they will slow down.

14. Discuss which of these conclusions are valid. [7]

Detecting tsunami waves

Tsunamis are large waves created by earthquakes or landslides that occur on the ocean floor. Because they carry a large amount of energy, they can cause large scale damage when they reach the shore and can be responsible for many deaths.

Countries that are at risk of tsunamis are keen to have early warnings of them. One way this can be achieved is by detecting seismic waves from the earthquakes and landslides that cause the tsunamis. The seismic waves can travel at speeds of 14,000 km/hr, while the tsunami itself is likely to travel at about 1,000 km/hr.

15. If an undersea earthquake occurred 7,000 km from land, calculate:

a) the length of time it would take the seismic wave to reach land

b) the length of time it would take the tsunami wave to reach land

c) the amount of warning time that people on the land would have. [4]

Another method of detecting tsunamis is to install ocean buoys. These have a pressure sensor at the bottom of the ocean, which can measure the depth of water above it. The buoy sits on the surface and relays the data from the pressure sensor to a satellite.

In deep water, tsunamis have a very long wavelength (about 100 km).

16. Calculate the amount of time a 100 km long wave would take to pass if it traveled at 1,000 km/hr. [2]

17. Identify an advantage and a disadvantage of each technique of detecting tsunamis. [4]

This tsunami warning buoy measures the ocean depth and can give valuable warning of a potential tsunami

6 Heat and light

Key concept: Systems

Related concepts: Environment, Development

Global context: Globalization and sustainability

▶ The heat from the Sun affects our environment in many ways. This picture of the Tatacoa Desert in Colombia shows a hot dry landscape, which is close to the equator. Other environments can be cold or wet. How do living things adapt to these different conditions?

▶ The way in which light behaves is important in astronomy, as it is the only way that we can work out what is going on in distant systems of stars. This picture shows a group of stars called the Pleiades. The light from these stars hits a cloud of dust, which causes the fuzzy appearance. Because blue light is scattered more easily than other colors, the fuzzy patches of light are bluer. A similar effect causes the sky to appear blue on Earth. How else can light carry important information?

Statement of inquiry:

Our environment is governed by the behavior of heat and light.

▲ It is easy to see that waterfalls transfer a large amount of energy. The Iguazu Falls on the border of Argentina and Brazil have about 1,750 tonnes of water fall over them every second. This water falls 60 to 80 m into the river below. The original source of the energy that is transferred in the waterfall is the Sun. We can convert some of this energy with hydroelectric power stations that use the energy of water moving downhill to provide electricity. How else do we capture the Sun's energy?

▶ Reflection occurs in nature, but only certain animals are able to understand that the reflected image is themselves, rather than another animal. The mirror test is used as a test of self-awareness. In the test, a small mark is painted onto the animal's face and the animal is shown a mirror. If the animal's response shows that it understands that the mark is on its own face (by rubbing it off, for example) then it has passed the test. Other than humans (which would only pass the test from the age of about 2) and apes, the only animals to have passed the test are dolphins, killer whales, magpies (pictured), an Asian elephant and a type of fish called the cleaner wrasse. Why do scientists try to assess the thought processes of animals and what are the difficulties in doing so?

Key concept: Systems

Related concepts: Environment, Development

Global context: Globalization and sustainability

Statement of inquiry:

Our environment is governed by the behavior of heat and light.

Introduction

In Chapter 4, Potential energy, kinetic energy and gravity, we saw that energy can be transferred from one form to another and that, in many situations, some energy was transferred to heat energy. In this chapter, we will see how heat and light can transfer energy.

The Sun is our most important source of heat and light. Without it, life on Earth would not be possible. The Sun's radiation controls the global environment, and variations in the Sun's radiation cause local environments and seasonal changes. In this chapter, we will see how the Sun's energy arrives on Earth and how it affects our environment. For this reason, one of the related concepts of the chapter is environments. The key concept of the chapter is systems because the weather patterns and climate systems are determined by the way in which heat energy is transferred.

Understanding how heat energy is transferred has allowed humans to improve our quality of life, particularly in environments with extreme weather conditions. For example, our understanding of insulation has allowed us to inhabit colder and more inhospitable regions of the Earth and develop more energy efficient buildings. The second related concept of the chapter is development and the global context is globalization and sustainability.

▼ Research taking place in Antarctica means that people live there all year round. This means that humans have been able to inhabit every continent on Earth

What is heat?

In Chapter 4, Potential energy, kinetic energy and gravity, we saw that heat is a type of energy. Chapter 8, States of matter, explains how the particles of a substance are continually moving around, and so every particle has kinetic energy. Although each particle is moving and has kinetic energy, the overall object is not moving and so will not have kinetic energy. Instead, the energy that it has, because of the motion of its individual particles, is called **thermal** or **heat energy**.

The amount of heat energy stored by an object could be increased by making its particles move or vibrate faster. The amount of energy stored could also be increased by having more particles—this means that the object has more mass. This is the reason that heavier objects take longer to heat up or cool down, as a larger amount of energy needs to be transferred.

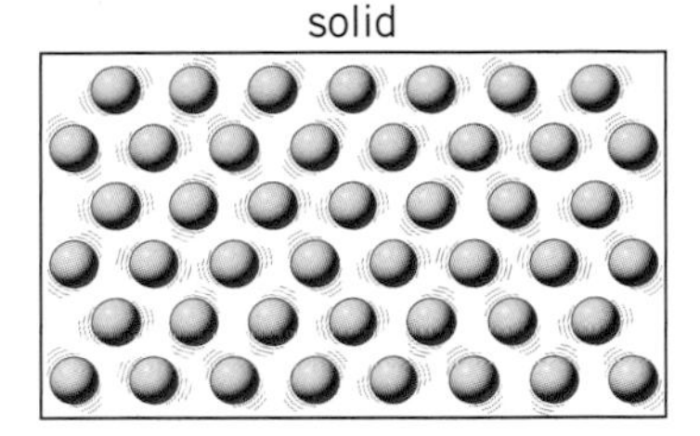

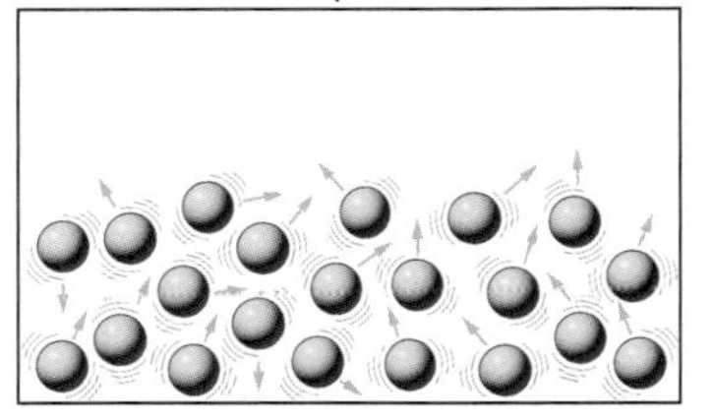

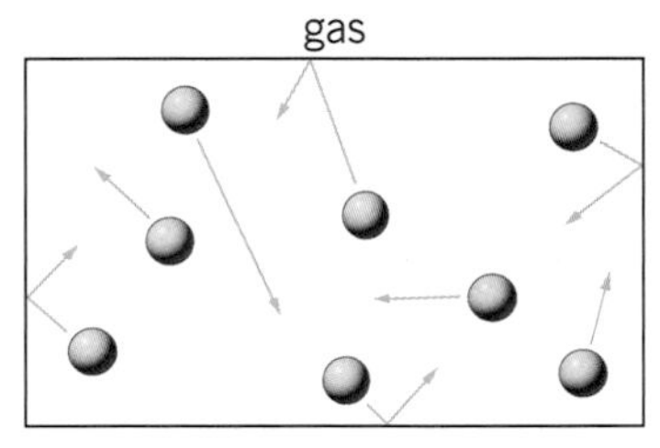

▲ The individual particles of solids, liquids and gases are continually moving and so have kinetic energy. Overall, however, the object is not moving and so the energy of the particles on a large scale is called thermal energy

▲ Because icebergs have a huge mass, it takes a large amount of thermal energy to melt them. This means that they last for a long time and can float large distances. The furthest that an iceberg from the Arctic has been known to drift is Bermuda

ABC **Thermal energy** is the energy that an object has through being hot, as a result of the motion of the object's particles. This is also called **heat energy**.

◀ This beach ball is filled with air. When the beach ball is hit, the whole ball and all the air molecules inside move together, and so the ball has kinetic energy. If the ball is left out in the sun, it will get hotter and the air molecules inside will move faster. The ball itself will not move as the molecules are moving in all directions, which cancel each other out. The beach ball has gained heat energy

Temperature is a measure of how hot something is. The temperature of an object is related to the average energy that each particle in the object has.

Degrees Celsius is a scale of temperature measurement where the boiling point of water is 100°C and the freezing point of water is 0°C.

Degrees Fahrenheit is a scale of temperature where water boils at 212°F and freezes at 32°F.

What is temperature?

Although heat energy is caused by the motion of tiny particles, it is impossible to measure this motion directly, in order to find the overall thermal energy of an object. However, **temperature** is a quantity that is much easier to measure. Temperature is a measure of the average energy possessed by each particle, and so the temperature of an object is directly related to the amount of heat energy it has.

Temperature scales

Many different scales of temperature have been used in the past. There are two that are still used widely in everyday life: **degrees Fahrenheit** (°F) and **degrees Celsius** (°C).

There is a formula that you can use to convert from one temperature scale to another. If F is the temperature measured in degrees Fahrenheit, and C is the temperature measured in degrees Celsius, then $F = \frac{9}{5}C + 32$ and $C = \frac{5}{9}(F - 32)$

1. The melting point of potassium metal is 63.5°C. What is this in degrees Fahrenheit?
2. On 10 July 1913, an air temperature of 134°F was recorded in Death Valley, California. This is thought to be the highest daytime temperature ever recorded. What is this temperature in degrees Celsius?
3. At what temperature is the measurement in degrees Celsius the same number as the measurement in degrees Fahrenheit?

Scientists use another scale of temperature: the Kelvin scale (K). If K is the temperature in Kelvin, then $K = 273.15 + C$. The Kelvin scale is important because 0 K (−273.15°C) is the lowest possible temperature, which is called absolute zero. At this temperature, the motion of the particles would stop altogether.

4. What is absolute zero in degrees Fahrenheit?

How can heat energy be transferred?

Like all types of energy, heat energy can be transferred from one object to another. It can also be converted to different forms, although this is sometimes not easy to achieve. Many processes lose energy to heat through processes such as friction in mechanical systems. It is impossible to recover all of this energy and transfer it all back to a more useful form. However, some of the heat energy can be transferred and used for other purposes.

◀ The electronics in this computer graphics card get hot because energy is wasted as heat energy. A fan is used to blow air over the circuit board. This transfers the heat energy away and stops the electronics getting too hot

What is conduction?

Conduction is a way in which heat energy can be transferred between two objects that are in contact. Where they touch, heat energy is transferred from the hotter object to the cooler object. The cool object gets hotter while the hot object gets cooler.

Heat energy can conduct along some materials more easily than others. Heat energy can be conducted through metals more easily than many other materials—they are called good thermal conductors. Other materials such as plastics are poor **thermal conductors**—these materials are called **thermal insulators**.

ABC

Conduction is a way in which heat energy is transported through a solid.

A **thermal conductor** is a material that allows heat energy to flow through it quickly.

A **thermal insulator** is a material that does not allow heat energy to flow through it well.

▼ Two ice cubes were put over a Bunsen burner. One was on a metal sheet (a good conductor), the other on an insulating heat mat. Because the metal is a good conductor, the ice cube on the metal melted quickly. The ice cube on the heat mat lasted for a much longer time

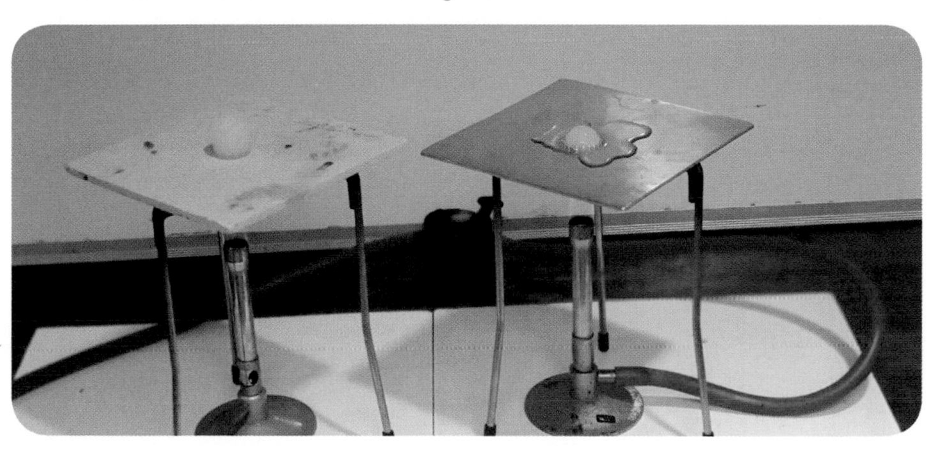

▶ The *Parker Solar Probe* is a NASA mission to send an unmanned spacecraft to the outer atmosphere of the Sun. On 29 October 2018, it passed just 25 million km from the Sun. At this distance, the temperature is very high and the probe is protected by a heat shield. This heat shield is only about 115 mm thick, but it can insulate the probe such that while temperatures on the Sun's side of the insulation are over 1,000°C, the probe will stay at about 30°C

Why does metal feel cold?

Metal often feels colder to the touch than other materials. This is because it is a good conductor: it allows the heat energy to flow away from your hand better than an insulator. It can also conduct heat energy to you quickly. This is why touching hot metal will be more painful than touching an insulator at a similar temperature.

To show this, find an insulating surface that is made from wood or plastic and a metal surface (such as a saucepan). Even if they are both at the same temperature, the metal surface will feel colder. However, if you put an ice cube on each surface, the one on the metal surface will melt faster. This shows that heat energy is conducted to the ice cube faster.

How fast does heat energy conduct?

ABC The **rate of conduction** is the speed at which heat energy is transferred from one object to another.

The **temperature gradient** is the difference in temperature between a hotter object and a colder object, divided by the distance between them.

The **rate of conduction** is the speed at which heat energy is transferred through an object from one side to another. There are various factors that affect the rate of conduction. Whether the material is a good thermal conductor or not is one factor and the **temperature gradient** is another. Conduction can also explain how heat energy is conducted from one object to another. In this case, the area of contact between the two objects will be important as well as the temperature gradient.

The temperature gradient is the difference in temperatures between the two objects or places that are transferring heat energy, divided by the distance between them. This can be expressed in the following equation:

$$\text{temperature gradient} = \frac{\text{temperature of object 1} - \text{temperature of object 2}}{\text{distance between object 1 and 2}}$$

The rate of conduction is directly proportional to this temperature gradient. This means that if the temperature gradient increases, the rate of conduction also increases.

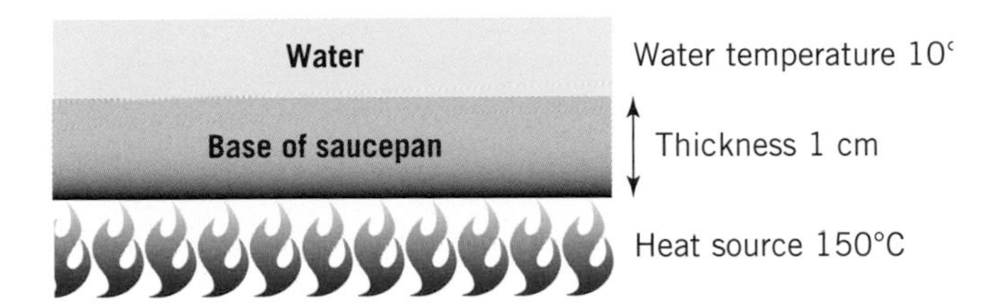

The diagram above shows the base of a saucepan on a heat source. In the diagram, the water in the saucepan has a temperature of 10°C, the heat source has a temperature of 150°C and the thickness of the saucepan base is 1 cm. The temperature gradient can be found using the equation

$$\text{temperature gradient} = \frac{\text{temperature of object 1} - \text{temperature of object 2}}{\text{distance between object 1 and 2}}$$

$$= \frac{150 - 10}{1} = 140°\text{C/cm}$$

What would happen to the temperature gradient as the water gets hotter?

1. A chef is stirring a pan of food that is hot. Explain why a wooden spoon might be better than a metal spoon for this.

2. A saucepan contains water at 20°C. If is put on a hot plate at 40°C, find the difference in temperature between the saucepan and the hot plate. Why would heat energy be transferred twice as quickly if the hot plate were at 60°C?

3. The wall of a house is 28 cm thick. The temperature inside the house is 19°C and outside it is 5°C.

a) Calculate the temperature gradient through the wall.

b) The house could be made warmer by having thicker walls. How else could the rate of conduction of heat energy through the walls be reduced?

▲ This woman has clothing that insulates her from the outside cold

Insulation

Insulating materials are often used in buildings to stop heat energy being conducted away. Good insulation will mean less energy is required to heat (or cool) a building. This will therefore reduce the costs of central heating (or air conditioning).

For this activity, you will need two beakers or cans and whatever insulating materials you can find. You might use cloth, newspaper or old packaging materials.

Insulate one of the beakers as well as your can. Don't forget to leave a lid or a gap to fill the beaker. Leave the other beaker uninsulated.

Fill both beakers with the same amount of hot water. Replace any lids and leave them for about 20 minutes. After this time, measure the temperature of the water in each beaker and compare your insulated beaker with the uninsulated one. Is your insulated beaker hotter?

Try to improve on your insulation. You could compete against classmates or compare your insulation to a vacuum flask.

Experiment

Measuring a cooling curve

Hot objects can lose heat faster by conduction and radiation. As a result, hot objects lose energy at a faster rate than colder objects.

Method

- Heat a beaker of water until the water is boiling. Alternatively, you could use a kettle to supply boiling water to put in a beaker.
- Remove the source of heat and measure the temperature of the water every minute using a thermometer.
- Record your data in a table.
- Plot your data as a graph of temperature in degrees Celsius (*y*-axis) against time in minutes (*x*-axis).

Questions

1. How can you tell how fast the water is losing heat energy from your graph?
2. Does your graph suggest that hot objects lose heat energy faster than colder objects?
3. How do you think your graph would have been different if you had used twice the volume of water?

ABC **Radiation** is the way that heat energy is transferred by hot objects emitting infrared waves.

Infrared waves are part of the electromagnetic spectrum. They are emitted by hot objects.

The **electromagnetic spectrum** is a spectrum of waves that includes visible light and infrared radiation, as well as radio waves, microwaves, ultraviolet, X-rays and gamma rays.

What is radiation?

Another way in which heat energy can be transferred is through **radiation**. Hot objects emit waves, which carry energy away. These waves are **infrared waves**. They are very similar to light: in fact, if an object is hot enough, it will glow and emit visible light as well as infrared radiation.

Because this thermal radiation is transferred as an **electromagnetic** wave, the energy can be transferred from one object to another without them being in contact. This energy can even travel through space—this is how we can feel the heat energy from the Sun.

1. Why can't heat energy get from the Sun to the Earth by conduction?

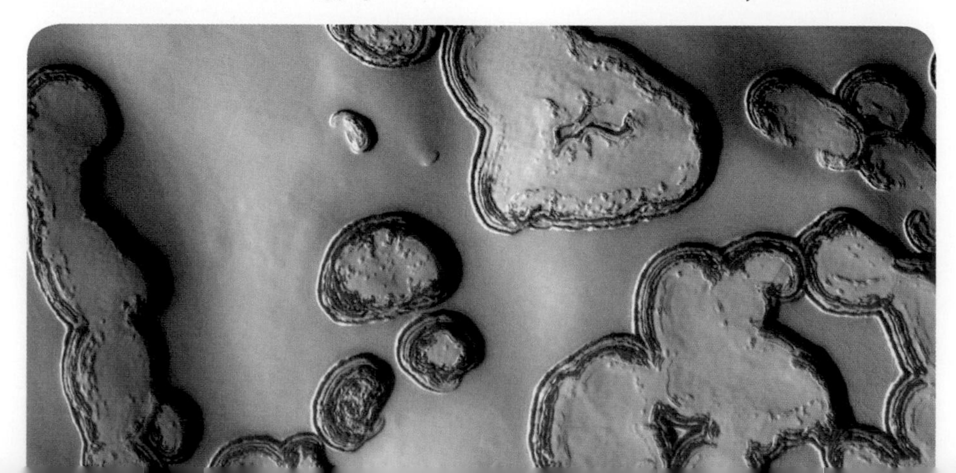

The Earth's environment is governed by the energy we get from the Sun. If the Earth were to be closer to the Sun, more of the radiation from the Sun would hit the Earth and the Earth would be much hotter. This picture shows the south pole of Mars, which is further from the Sun than the Earth. As a result, Mars is much colder. Although there is not much water present on Mars, ice forms at the poles. The ice shown here is solid carbon dioxide ▶

What affects the rate of radiation?

The first factor that affects the rate of radiation that is emitted is the temperature. Objects with a higher temperature radiate more than things with a lower temperature. An infrared camera can detect this radiation. These cameras can therefore be used to see hot objects, even in the dark. For this reason, infrared cameras are used for night vision, as well as for monitoring the temperature of things.

Another factor that affects the amount of radiation that is emitted is the color of the surface. A white surface does not emit as much radiation as a black surface.

The color of the surface also affects how the radiated heat is absorbed. Light surfaces reflect most of the radiation, but dark surfaces absorb heat energy more easily. Therefore dark surfaces heat up more quickly than light surfaces.

▲ This infrared picture shows a person holding a cup of tea. The hot cup radiates much more energy than everything else in the picture

▲ This metal box is filled with boiling water and therefore the inside of the box is all at the same temperature. However, the metal box has different colored surfaces. The left-hand surface is black and therefore radiates much more infrared radiation than the right-hand side, which is silvery. This is particularly obvious when viewed with an infrared camera. The black surface appears to be at a temperature of 88°C, while the silvery surface is at a temperature of 44°C

◀ Snow is white, which means it reflects most of the infrared radiation from the Sun. As a result, it does not absorb much energy and it stays cold. If the snow gets slushy and darker, or is next to a darker object, it absorbs energy at a higher rate and will melt more quickly. Dry snow is also a poor conductor of heat. Shelters built from snow, or igloos, are effective at insulating people inside from the cold weather outside

Snow and the Earth's climate

The amount of snow covering the Earth affects how much of the Sun's energy is absorbed and how much is reflected. This in turn affects the Earth's climate. The global snow lab at Rutgers University has been measuring the amount of land in the northern hemisphere that is covered by snow each week. A graph of their data from 2000 onwards is shown below.

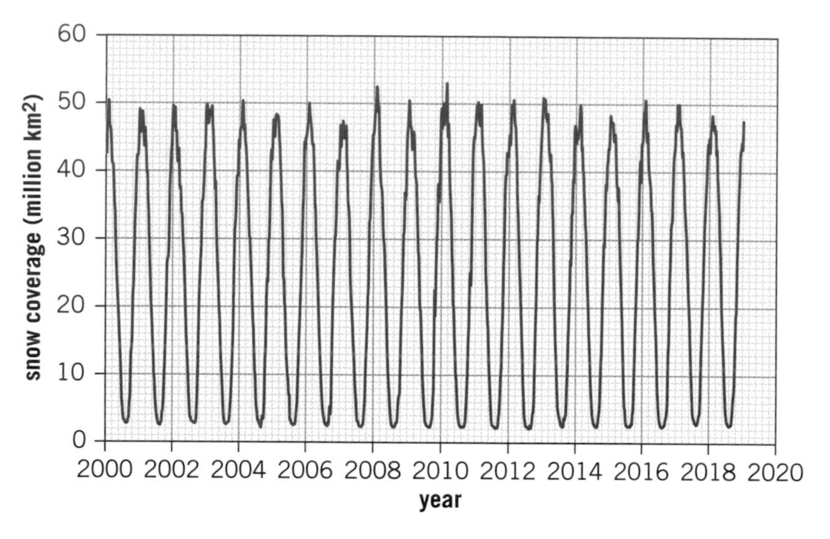

1. When was the highest amount of snow coverage recorded?
2. Use the graph to estimate the area of land in the northern hemisphere that is permanently covered in snow.
3. The northern hemisphere has a land area of approximately 100 million km^2. Use the graph to estimate the maximum percentage of land in the northern hemisphere that was covered by snow in the winter of 2017–2018.
4. Does this graph show any evidence of global warming?
5. The southern hemisphere has more ocean than the northern hemisphere and only about 50 million km^2 of land. 14 million km^2 of this is in Antarctica, which is almost entirely covered in snow for the whole year. How might a graph of the snow coverage of the southern hemisphere differ from the graph of the northern hemisphere above?

ABC

Visible light is a type of electromagnetic wave that our eyes are able to detect.

Microwaves are a type of electromagnetic wave with wavelengths of between 1 mm and 1 m. They are used for some communications and for heating food in microwave ovens.

What is light?

We have seen that hot objects can emit thermal (infrared) radiation. If something is hot enough (more than about 525°C) then it will start to glow and emit **visible light**. This is because light and infrared radiation are both part of the electromagnetic spectrum. This also includes **microwaves**, **radio waves** and **X-rays**. These different types of waves in the spectrum will have different wavelengths.

The visible light that our eyes can see is only a very small part of this electromagnetic spectrum. However, it is a very important part of the spectrum as it allows you to see.

Radio waves are a type of electromagnetic wave with a long wavelength. They are used for communication.

X-rays are a type of electromagnetic wave with a very short wavelength. X-rays are able to pass through matter. This property means that they are useful for medical imaging.

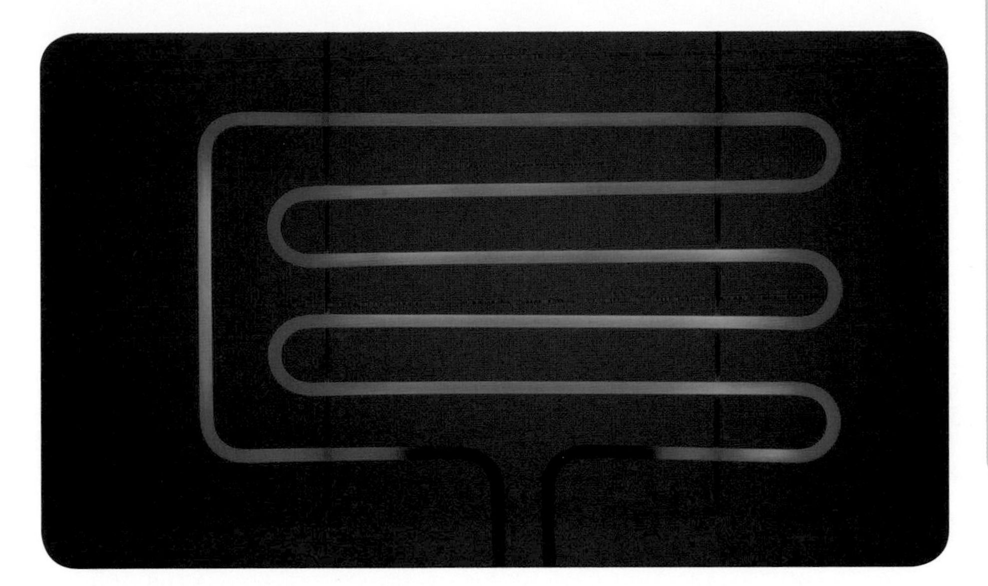

▲ This heating element is glowing. It is emitting thermal radiation as well as red light. This is because visible light and infrared radiation are all part of the same electromagnetic spectrum

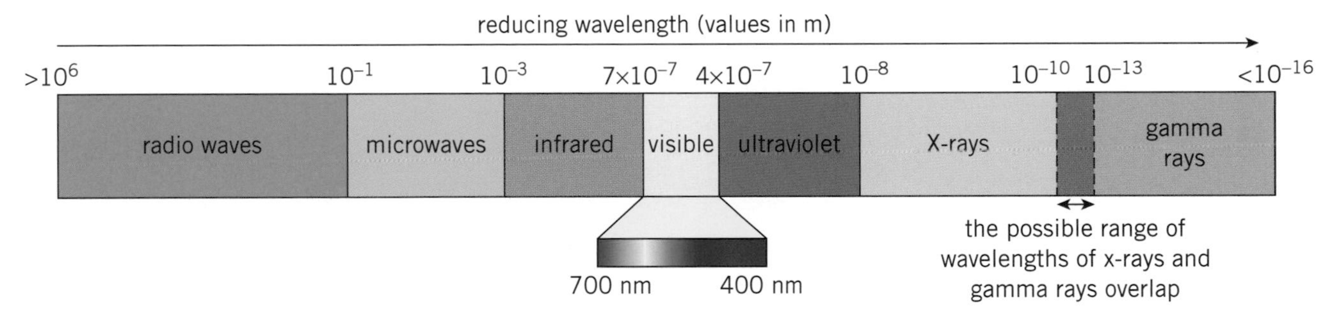

▲ This diagram shows the electromagnetic spectrum, which is a set of waves that behave in similar ways. Each type of wave is defined by a range of wavelengths. For example, the electromagnetic wave with a wavelength between 700 nanometers and 400 nanometers is visible light, as shown here. What is the range of wavelengths for microwaves?

How does light behave?

Light tends to behave in certain ways. It travels so fast that we see it instantly, it travels in straight lines and it can be absorbed or reflected from things. When light behaves differently, our brains are tricked by what our eyes see. For example, a mirror can persuade our brains that the reflections that we see are real objects.

▶ Because light normally travels in straight lines, the cat's brain assumes that there is another cat behind the mirror

A **reflection** occurs when a wave bounces off a surface.

The **angle of reflection** is the angle between the normal and the reflected ray.

The **angle of incidence** is the angle at which a ray of light hits a surface. The angle is measured to the normal.

The **law of reflection** states that the angle of incidence is equal to the angle of reflection.

The **normal** is a line that is at a right angle to the surface.

A **specular reflection** is a reflection off a shiny surface in which a reflected image can be seen.

A **diffuse reflection** is a reflection of light off a surface that is not shiny enough for a reflected image to be seen.

How does light reflect?

When light hits a surface, it can be **reflected**. This means that the light bounces off the surface at the same angle as it hits the surface.

- The angle at which the light hits the reflective surface is called the **angle of incidence** (*i*).
- The angle at which the light reflects is called the **angle of reflection** (*r*).

Both these angles are measured to the **normal**. This is a line that is at a right angle to the surface. This is summarized in the **law of reflection**, which states that the angle of incidence is equal to the angle of reflection.

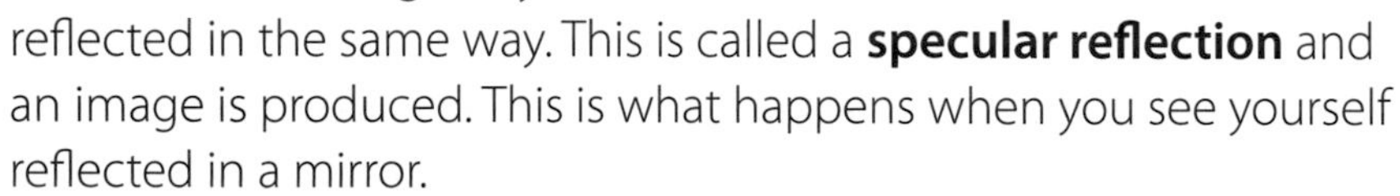

When light hits a flat, shiny surface, all of the light rays are reflected in the same way. This is called a **specular reflection** and an image is produced. This is what happens when you see yourself reflected in a mirror.

Sometimes a surface is not flat. In this case, the angle between the normal and incident light will vary at different points on the surface and the angle of reflection will also vary. This means that the light will be scattered in many different directions and no image will be seen. This is called a **diffuse reflection**.

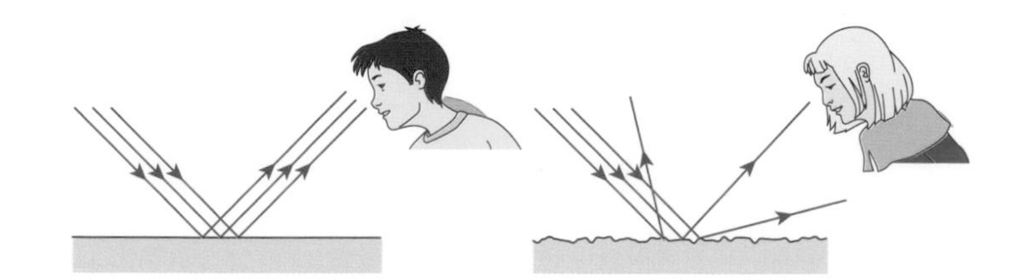

▶ When light hits a flat surface, it is reflected at the same angle at which it hit the surface and an image can be seen. This is a specular reflection. When light hits a rough surface, it hits the surface at different angles and so each ray of light is reflected at a different angle. This is called a diffuse reflection, and no image is seen

1. A ray of light hits a pair of mirrors that are positioned at right angles.

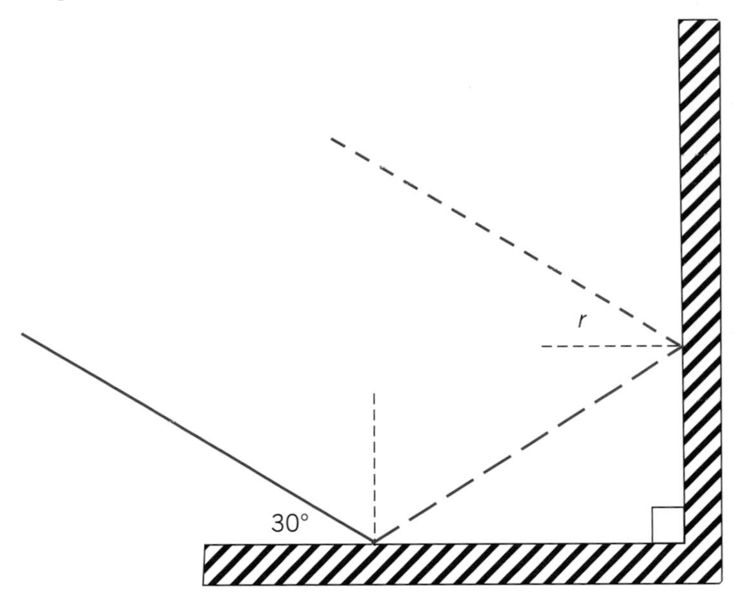

a) What is the angle of incidence of the light ray?

b) What is the angle of reflection of the light ray from the first mirror?

c) Find the angle marked *r*.

d) If the original angle of the ray, as marked in the diagram, were increased, what would happen to the angle *r*?

When is light reflected?

We saw earlier that lightly colored surfaces tend to reflect heat while dark surfaces absorb heat. The same is true for light. Dark surfaces appear dark because they reflect less of the light that hits them. Light surfaces, on the other hand, reflect more of the light and appear lighter. If the light is absorbed, then the material will absorb the energy that the light carries and so will get hotter. This is why a black surface in the sun will get hotter than a white surface.

Experiment

Do black surfaces absorb heat better than white surfaces?

Method

- Place a 2 cm depth of water in two transparent bowls or beakers.
- Measure the initial temperature of the water.
- Place one bowl on a piece of black paper and the other on white paper.
- Place both bowls in the Sun and leave for an hour.
- Measure the temperature of the water.

Questions

1. Does the experiment suggest that black surfaces absorb heat energy better than white surfaces?
2. What were the control variables in this experiment?
3. Were there any variables that should have been controlled but were not?

ABC A **rainbow** is a naturally occurring effect where sunlight is split into a spectrum of colors.

The **visible light spectrum** is a spread of colors. The order of these colors in the spectrum is red, orange, yellow, green, blue, indigo and violet. These are also known as the **spectral colors**.

How can light appear as different colors?

Sometimes the light from the Sun is split into a **rainbow**. This shows that the white light is in fact a combination of all the colors of the rainbow mixed together. When it is sunny and raining at the same time, the light from the Sun will be split by the raindrops into its full **spectrum** of colors. A similar effect can be achieved by shining a light into a prism.

The colors in the spectrum will always appear in the same order: red, orange, yellow, green, blue, indigo, violet. This is the order of the colors from high to low wavelength.

How do different surfaces appear different colors?

Different surfaces will reflect some colors of light but absorb others. The colors of light that are reflected will determine the color that the surface appears to be. For example, a blue surface appears blue because it reflects blue light but absorbs other colors of light. If a blue object was viewed under red light, the light would be absorbed, and the object would appear black.

▼ Sometimes, the water droplets in the atmosphere can cause the Sun's light to be split into its spectrum of colors. The result is a rainbow

▲ Most plants have leaves that appear green. This is because chlorophyll, the chemical that is responsible for photosynthesis, absorbs red and blue light. This light can be converted into useful forms for the plant. The green light, however, is reflected. When plants are grown under artificial light, they need blue and red light

What are primary and secondary colors?

Our eyes can only detect three colors: red, green and blue. We call these the **primary colors.** Other colors that we see can be made from mixtures of the primary colors.

Secondary colors are equal mixtures of two of the primary colors. Magenta is a mixture of blue and red, cyan is blue and green, and yellow is red and green.

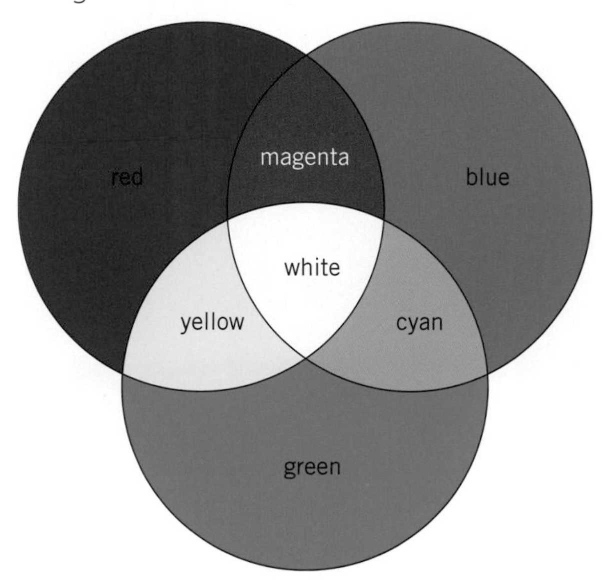

▲ The primary and secondary colors

ABC The **primary colors** are the colors that can be detected by the human eye. They are red, green and blue.

The **secondary colors** are equal mixtures of two of the primary colors. The secondary colors are cyan (blue + green), magenta (blue + red) and yellow (green + red).

Colored flags

Many flags use primary colors. For example, the American flag uses red, white and blue. If the flag were to be viewed under red light, the red stripes would still appear red. However, the white stars and stripes would also appear red, since the white parts of the flag would reflect the red light. The blue part would appear black, as blue surfaces absorb all colors of light other than blue.

▲ The American flag as viewed under white light and how it would appear if it were viewed under red light.

1. How would the American flag appear under these colors of light?

   a) Green

   b) Blue

   c) Magenta

▲ When viewed close-up, a computer screen only has three different colors. This is because our eyes only detect red, green and blue. We see other colors as combinations of these three colors

How is the Sun's heat and light important to us?

The Sun's heat and light have a large effect on our environment. At the equator of the Earth, the intensity of the Sun is always strong, and places near the equator are hot. At higher latitudes, nearer to the poles of the Earth, the Sun is low in the sky in winter and the days are short. As a result, the winters are colder in these places.

▲ The different intensity of the Sun at different places on Earth creates very different environments. These pictures show Antarctica, and the Thar Desert in India

ATL Self-management skills

Emotional management

Our environment has a very important impact on our emotional wellbeing. Factors such as weather and daylight can have a large impact on our moods. In countries where there is a larger seasonal variation in weather and fewer hours of daylight in winter, people's moods can be affected. This seasonal change is normal; however, some people can suffer from depression during winter. This is called Seasonal Affective Disorder (SAD).

Some people can manage SAD and reduce its effects with a diet that is high in vitamin D (fish for example) and outdoor exercise in daylight hours. Others use light boxes to simulate daylight.

▲ How does the weather influence your mood?

How do we use the Sun's energy?

The Sun is transferring large amounts of energy to us all of the time. Plants have evolved to capture this energy and they use it to grow. In turn, animals eat plants (and each other) and the energy that originally came from the Sun is passed on.

Humans use energy in other ways. We heat our homes and we use electricity to supply energy to different appliances. This means that we have need for energy resources other than the food that we eat. Almost all of this energy still comes from the Sun.

Fossil fuels such as coal, oil and natural gas are the remnants of plants that decomposed millions of years ago. Energy from the Sun was captured by these plants and stored as chemical energy. Burning the fossil fuels releases this energy. It can then be used as heat energy to heat our homes, or a power station can transfer the heat energy to generate electricity.

However, the creation of fossil fuels is a natural process that takes millions of years and one day supplies may run out as we burn more of them. Burning fossil fuels also releases **greenhouse gases** that may prove to change the global climate in a catastrophic way. For these reasons, more countries are trying to get their energy requirements from **renewable sources**. This means making more use of the large amounts of energy that the Sun transfers to the Earth every day.

1. Burning fossil fuels uses this finite resource as well as releasing greenhouse gases. Despite this, fossil fuels are used to provide more than half of the world's electricity.

   a) What advantages do fossil fuels have over other ways of generating electricity?

   b) Other than generating electricity, what other uses of fossil fuels are there?

One way in which the Sun's energy may be used is to capture the heat or light energy from the Sun directly. **Photovoltaic cells**, for example, convert the Sun's light energy into electricity. Photovoltaic cells can be used on a small scale, where an individual household supplies some or all of its electricity from a photovoltaic panel. Alternatively, large fields with many photovoltaic cells can be constructed to generate larger amounts of electricity. These are known as solar farms. The heat of the Sun can also be used to supply hot water to a house and sometimes it can be used to generate electricity as well.

ABC

Fossil fuels are the remains of animals and plants that have been converted into coal, oil or gas over millions of years.

Greenhouse gases add to the greenhouse effect and cause the Earth's climate to get warmer.

Renewable energy sources are sources of energy that will not run out.

A **photovoltaic cell** converts the Sun's light energy into electricity.

▶ These photovoltaic cells convert the Sun's light into electricity, which can be used to supply the house

▶ Solar farms use large fields of photovoltaic panels to generate larger amounts of electrical power

ABC **Hydroelectric power** is a way of generating electrical energy from the kinetic energy of the flow of water.

The energy from the Sun also drives our weather systems. The heat of the Sun causes water from the oceans to evaporate, which later falls as rain. Some of this rain falls on higher ground and then runs downhill in rivers. **Hydroelectric power** uses the kinetic energy of water flowing down hills to generate electricity. If a river is dammed, then water can be stored at the top of the dam in a lake. This creates a huge pressure of water. When the water is allowed to flow through the dam, it can drive a turbine and generate electricity.

▲ This hydroelectric power station in Arizona has a dam that is 220 m high. Allowing water to flow from the lake at the top powers a turbine and generates electricity

Changes in temperature on Earth cause wind to blow. This could be the change in temperature between day and night, or between different places. The kinetic energy of the wind's motion can be transferred into electrical energy using a **wind turbine**. This can provide a useful source of energy in places that are reliably windy.

ABC A **wind turbine** is a way of generating electrical energy from the kinetic energy of the wind.

◀ Wind farms convert the kinetic energy of the wind into electrical power

Solar power, hydroelectric power and wind power are all renewable sources of energy because they use energy from the Sun that has not been stored. These sources of energy will not run out as long as the Sun continues to shine. They are also considered clean sources of energy because they do not emit greenhouse gases.

However, there are problems associated with these sources of energy. They can cost a lot of money to install and may require the use of large areas of land. They can also be unreliable because the weather can change and the amount of sunlight, wind or rain is variable.

1. Wind, hydroelectric and solar power are all renewable sources of electricity.
    a) Give an advantage of each method of electricity generation.
    b) Give one disadvantage of each method of electricity generation.
    c) Imagine that you are tasked with helping your school to source as much of its electricity from renewable sources. Which of these three renewable sources would be most useful in your location? How could they be installed?

Energy from the Sun

The graph below shows the Sun's energy that hits the Earth in any day of the year. The x-axis is the day of the year counted from 1 January. The amount of energy depends on the latitude.

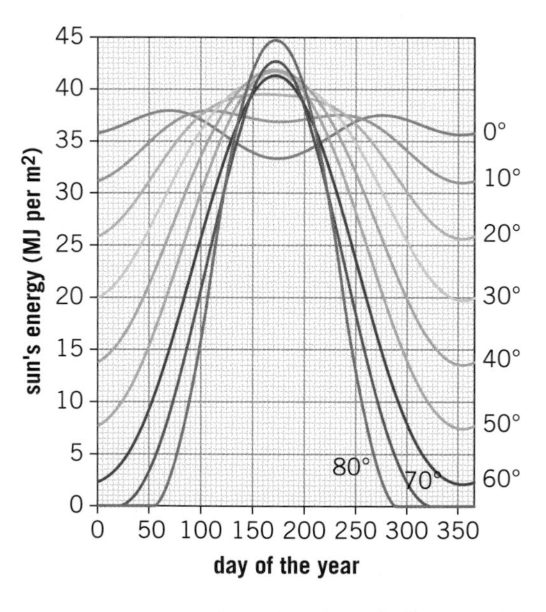

1. Find out the latitude of the place where you live.
2. Use the graph to find out how much energy from the Sun will hit the ground today in your location.
3. Find the maximum and minimum energy from the Sun in your location.

Summative assessment

Statement of inquiry:

Our environment is governed by the behavior of heat and light.

We use a lot of energy in our lives and much of this comes from fossil fuels. In this assessment we look at how we can harness the energy from the Sun and how saving energy by using insulation can help reduce the amount of energy that we use.

The behavior of heat and light

1. Which is the best thermal conductor?
   - **A.** Metal
   - **B.** Plastic
   - **C.** Water
   - **D.** Wood
2. Which of the following is not a primary color?
   - **A.** Blue
   - **B.** Cyan
   - **C.** Green
   - **D.** Red
3. What is the best explanation for why grass is green?
   - **A.** It radiates green light but not red or blue light
   - **B.** It absorbs green light but not red or blue light
   - **C.** It reflects green light but not red or blue light
   - **D.** It glows with a green light but not with a red or blue light
4. What is the law of reflection?
   - **A.** The angle of incidence is equal to the angle of reflection.
   - **B.** The colors of light are reflected equally.
   - **C.** Light is always reflected.
   - **D.** Different colors of light are reflected through different angles.
5. If you warm your hands by a fire, the heat energy reaches you by
   - **A.** Conduction
   - **B.** Radiation
   - **C.** Reflection
   - **D.** Light energy

Investigating how the Sun's energy is absorbed

A pupil wants to measure how different colors absorb different amounts of radiation. He writes the following method:

- Take a beaker with 100 cm^3 of water. Wrap it in blue paper.
- Measure the temperature of the water.
- Leave the beaker of water in the Sun for one hour.
- Measure the final temperature of the water.
- Now replace the blue paper with red paper. Measure the temperature of the water and leave the beaker in the Sun for one hour. Measure the final temperature of the water.
- Repeat the experiment for green, white and black paper.
- Plot a graph of the results.

6. Write a suitable hypothesis for this experiment. [2]
7. What type of graph should the pupil plot? [1]
8. What are the independent and dependent variables of this experiment? [2]
9. Identify one control variable from the method. [1]
10. Give the name of one piece of equipment that the pupil will need to make a measurement. [1]
11. Identify one problem with this experiment and suggest an improvement to the method. [3]

Investigating insulation

Insulating buildings and our homes can help to reduce the amount of energy that is used to heat them.

In an experiment to investigate insulation, a beaker with a lid was wrapped in sheets of newspaper. The beaker was filled with boiling water and left for 20 minutes. After this time, the beaker was unwrapped and the temperature of the water was measured.

The results of this experiment are shown below.

Number of sheets of newspaper	Final temperature (°C)	Temperature change (°C)
0	46	54
2	50	
4	57	
6	60	
8	65	
10	66	34

12. Explain why it was important to use the same volume of boiling water each time. [2]
13. The temperature change was calculated by assuming that the water started at 100°C. On a copy of the table, fill in the missing values of the temperature change. [2]
14. Plot a graph of temperature change (*y*-axis) against number of sheets of newspaper (*x*-axis). [3]
15. Add a line of best fit to your data. [1]
16. Use your graph to estimate the temperature change that would occur with five sheets of newspaper insulating the beaker. [2]

Solar farms

Photovoltaic cells can be used to make a solar farm. This is where many photovoltaic panels are placed in a field and used to generate electricity from the Sun's light.

17. Imagine that you are a farmer who is considering converting some of his fields into a solar farm. Identify and explain one benefit of doing this and one problem of the plan. [4]

18. The farmer decides to go ahead with building a solar farm. Write a letter to the residents of the local town to persuade them that this is a good idea. In your letter, you should explain the energy transfers involved in a photovoltaic panel and describe some advantages of using these to generate electricity over alternative methods. [6]

▲ A solar farm

7 Atoms, elements and compounds

Key concept: Systems

Related concepts: Patterns, Models

Global context: Identities and relationships

◀ We use different elements and materials in industry and the source of these elements is from the Earth. Some elements are abundant and easily available. Other elements, such as gold, are rare. Large mines such as this have to be constructed to find gold. Which other important materials do we dig from the Earth?

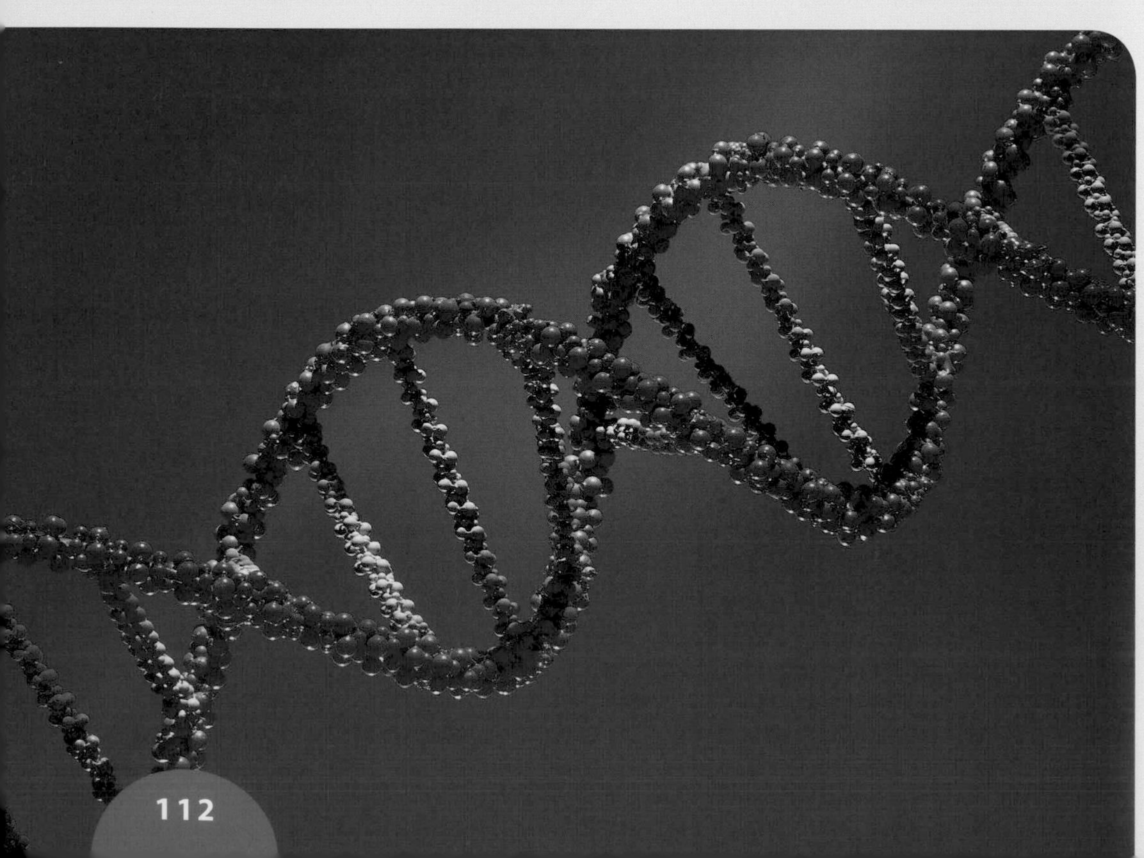

◀ Atoms can combine to form complicated compounds. Our genetic identity is stored in a DNA molecule, which contains hundreds of billions of atoms. The structure of DNA is complex, but it is made from only a few different types of atom. Where else can complicated patterns arise from a simple starting point?

Statement of inquiry:

The complex chemicals that enable life to exist are formed from only a few different types of atom.

▲ This picture shows the surface of Mars. The same elements that are present on the Earth are also present on Mars. For various reasons, they have not combined in the same ways to create the life that we see on Earth. Which factors might be important for life to exist?

◀ These nylon fibers are being woven into cloth. Nylon, like many other plastics, is a long molecule consisting of carbon, nitrogen, oxygen and hydrogen atoms. The first type of nylon was invented by the American chemist, Wallace Carothers, in the 1930s. Its use in clothes, fabrics, ropes and many other things has made it a popular material. What are the drawbacks of using plastics?

Key concept: Systems

Related concepts: Patterns, Models

Global context: Identities and relationships

Statement of inquiry:

The complex chemicals that enable life to exist are formed from only a few different types of atom.

Introduction

The matter that makes up our bodies, the Earth, plants and animals, and the buildings in which we live and work, is made of atoms. However, the atoms are very small, so it is hard to tell that they are there. An adult human being has about 10 thousand million million million million atoms in their body. About 98% of these atoms are one of only three types: hydrogen, oxygen and carbon. These three types of atom make up about two thirds of our body mass. This small set of building blocks can combine to form all sorts of complex structures and materials that make up all life on Earth. Small changes in our genes are caused by a small change in the atoms of our DNA. These small changes mean that every person is different, with a unique individual identity. The global context of this chapter is identities and relationships.

In this chapter, we will look at what these atoms are and how they can mix and combine to form the matter around us. We will look at some of the systems for naming and classifying this matter and therefore the key concept is systems. Patterns in the structure of atoms can be used to organize elements into groups. This is how the periodic table was created. We will investigate the models used to understand how atoms work. As a result, the related concepts are patterns and models.

◀ Around 3 billion years ago, some bacteria evolved to develop a method of capturing the Sun's energy, converting it into a chemical store. This was the earliest form of photosynthesis and one of the most important events in the evolution of life on Earth. Today, plants use photosynthesis to create carbohydrate molecules. In doing so, the plant releases oxygen. This process provides food for the plant and other animals, as well as providing the oxygen in our atmosphere

People are all different. However, the chemical substances that we are made from are almost identical. We are all made of atoms. Most of these atoms just three types: hydrogen, oxygen and carbon

What are atoms?

Atoms were first thought of by the ancient Greeks more than 2,000 years ago. They wondered what matter was made from and imagined what the smallest possible unit of matter could be. At the time, they thought that all matter was made of four elements: air, earth, fire and water.

In 1803, John Dalton, an English chemist, published a new theory of how matter was made from atoms. His theory was based on the observation that elements combined in simple ratios to form other chemical compounds. The atomic theory stated that atoms cannot be created, destroyed or divided (although we now know that in high-energy physics experiments, this is not true). This means that the total mass in a chemical reaction is conserved.

For example, when methane gas (the most common constituent of natural gas) is burned in oxygen, 4 g of methane will require 16 g of oxygen to burn completely. The reaction will release 11 g of carbon dioxide and 9 g of water. As a result, the original mass of gas ($4 + 16 = 20$) is the same as the mass of the products of the reaction ($11 + 9 = 20$).

1. When 36 g of water is split into its constituent elements, hydrogen and oxygen gases are released. 4 g of hydrogen gas is formed. Using the atomic theory, determine what mass of oxygen gas is released.

Experiment

Comparing ratios of reacting quantities

Calcium carbonate is a chemical compound found in limestone. It decomposes on heating to form a substance called quicklime, or calcium oxide. Carbon dioxide is also given off in this reaction. The equation for this reaction is $CaCO_3 \rightarrow CaO + CO_2$. In this experiment you will compare the mass of calcium carbonate to the mass of calcium oxide that is formed.

Safety

- Wear safety glasses at all times.
- Wear safety gloves when handling calcium carbonate.

Method

- Measure the mass of an empty crucible.
- Place two spatulas of calcium carbonate in the crucible. Measure the new mass of the crucible with the calcium carbonate in it. Using the mass you recorded for the empty crucible, calculate the mass of calcium carbonate in your crucible. If this is done as a class experiment, your teacher may provide each group with a different mass of calcium carbonate at the start.
- Heat the crucible using a Bunsen burner. You should heat it for about five to ten minutes.
- Turn off the heat and allow the crucible to cool.
- Once the crucible is cool, measure the mass of the crucible with the newly formed calcium oxide inside.

Questions

1. Calculate the mass of calcium oxide.
2. Calculate the mass of carbon dioxide released.
3. Calculate the mass of carbon dioxide as a percentage of the original mass. The percentage of carbon dioxide can be calculated using the formula

$$\frac{\text{mass of carbon dioxide}}{\text{initial mass of calcium carbonate}} \times 100\ \%$$

4. Compare the value of your percentage with other members of your class.
5. It is possible that not all the calcium carbonate reacted. How would this affect the value of your percentage?

In this picture, an electron microscope is used to detect individual atoms. Here, a single atom of silicon shows up as an impurity in graphene. Graphene is a structure containing a layer of carbon atoms

What is an atom made of?

At first, atoms were thought to be the smallest possible components of matter. However, in 1896, it was discovered that atoms can decay from one element into another. This process, known as radioactivity, suggested that atoms must be made of even smaller parts. Further experiments at the turn of the 20th century showed that atoms were made of **electrons**, **protons** and **neutrons**. We call these subatomic particles. The properties of these subatomic particles are summarized in the table below:

Subatomic particle	Relative charge	Relative mass
electron	−1	0.00054
proton	+1	1
neutron	0	1

Protons and neutrons have essentially the same mass, while electrons are almost 2,000 times lighter. The subatomic particles are also defined by their **charge**. Protons are positively charged, electrons have a negative charge and neutrons are **neutral**. The protons and neutrons are found at the center of the atom, called the **nucleus**. The electrons orbit around this nucleus.

A **proton** is a positively charged particle found in the nucleus of the atom.

A **neutron** is a neutral particle with the same mass as a proton found in the nucleus of the atom.

An **electron** is a negatively charged particle that orbits the nucleus in the atom.

Charge is a property of matter that is responsible for electric fields and electricity. Objects can be positively charged, negatively charged, or neutral.

If a particle is **neutral**, then it has no charge.

The **nucleus** is the center of an atom, and contains the protons and neutrons. The plural of nucleus is nuclei.

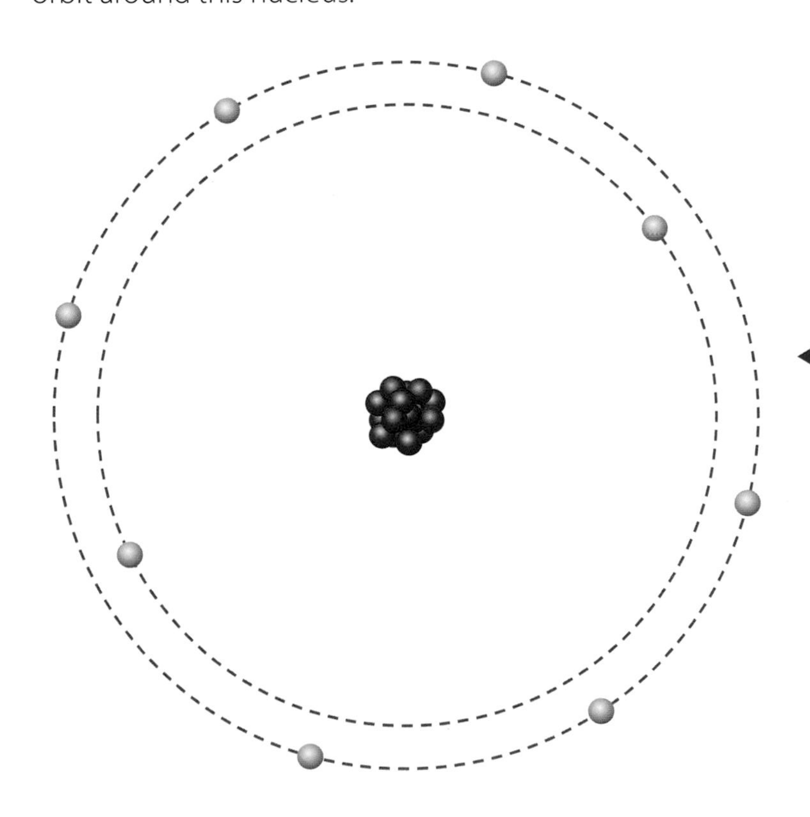

The protons and neutrons are found in the nucleus of the atom and the electrons orbit around the outside. In this diagram of an oxygen atom, there are 8 protons, 8 neutrons and 8 electrons

The atom itself is neutral. This means that the number of protons is equal to the number of electrons—the positive charge of the protons is canceled out by the same amount of negative charge from the electrons.

Although the protons and neutrons are much heavier than the electrons, the amount of space they occupy is tiny. The electrons occupy the entire size of the atom (about 10^{-10} m) and the nucleus is about 100 thousand times smaller. If the atom were the size of a large stadium with the electrons orbiting in the stands, the nucleus would be the size of a pea placed in the middle of the stadium.

How can atoms differ?

ABC The **atomic number** is the number of protons in the nucleus of an atom.

An **element** is matter that consists of only one type of atom.

Because the nucleus is so small and tucked away in the center of the atom, it very rarely interacts with anything outside the atom. As a result, the properties of an atom are determined by the electrons. For an atom to be neutral, it must have the same number of protons and electrons. Therefore, the number of protons is the same as the number of electrons. The number of protons in the nucleus is often referred to as the **atomic number**, because it defines which type of atom it is. Different types of atom are called **elements**.

Atomic number (proton number)	Element
1	hydrogen
2	helium
3	lithium
4	beryllium
5	boron
6	carbon
7	nitrogen
8	oxygen
9	fluorine
10	neon

▲ The first ten elements are defined by the atomic number—the number of protons in the nucleus

▲ Carbon can exist as graphite, a black powder, or diamond, a hard crystal. Atoms of carbon have an atomic number of 6, one less than nitrogen. However, despite a small change in atomic number, nitrogen and carbon have very different properties

Each different element has different chemical properties. For example, carbon has six protons and exists as a solid in the form of graphite or diamond. Nitrogen has seven protons, which means that it has one more electron than carbon. Despite this seemingly small difference, nitrogen exists as a gas and forms most of the air around you. Solids and gases are explored in Chapter 8, Properties of matter.

Elements can also differ in their overall mass. The numbers of protons and neutrons in the nucleus determine the mass of an atom—the mass of the electrons is negligibly small. The protons and neutrons have almost identical masses.

Adding together the total number of protons and neutrons in an atom gives the **mass number**.

1. If you know the mass number and atomic number of an atom, how might you calculate its number of neutrons?

Sometimes two atoms can have the same number of protons but different numbers of neutrons. As a result, their atomic numbers are the same, but their mass numbers are different. Such atoms are called **isotopes**. Because the atoms have the same number of protons, they are the same type of element and have the same chemical properties, but the difference in the number of neutrons gives the atoms different masses.

ABC The **mass number** is the number of protons and neutrons in the nucleus of an atom.

Isotopes are two or more atoms with the same number of protons in their nuclei (and hence the same atomic number) but a different number of neutrons (hence a different mass number).

Atomic number (proton number)	Mass number of most common isotope	Element
1	1	hydrogen
2	4	helium
3	7	lithium
4	9	beryllium
5	11	boron
6	12	carbon
7	14	nitrogen
8	16	oxygen
9	19	fluorine
10	20	neon

▲ The first ten elements with the mass number of the most common isotope

2. Use the table showing the atomic numbers and mass numbers of the first ten elements to answer these questions:

a) Which two elements have ten neutrons in the nucleus of their most common isotope?

b) Which elements have the same number of protons as neutrons in the nuclei of the most common isotope?

c) Which element has fewer neutrons than protons in the nucleus of the most common isotope?

Normally, an element has one isotope that is far more common than other isotopes. For example, the vast majority of all oxygen atoms on Earth (99.76%) have eight protons and eight neutrons, which gives the oxygen a mass number of 16. This isotope is therefore known as oxygen-16.

However, there are very few oxygen atoms that have an extra neutron (0.04%), giving these atoms a mass number of 17, or even 2 extra neutrons (0.2%) and a mass number of 18. These isotopes are known as oxygen-17 and oxygen-18.

Isotope	Atomic number	Number of neutrons	Mass number	Abundance on Earth
oxygen-16	8	8	16	99.76%
oxygen-17	8	9	17	0.04%
oxygen-18	8	10	18	0.2%

ABC The **relative atomic mass** is the average mass of an atom of that particular element. The mass is relative to carbon-12, which is defined to have a relative atomic mass of 12.

The mass number is only an approximation for the mass of an atom. The presence of different isotopes affects the average mass of an atom of any given element and a more precise measure of the average mass is sometimes required. This is called the **relative atomic mass**.

Normally, the relative atomic mass is very similar to the mass number of the most abundant isotope. For example, the relative atomic mass of oxygen is 15.999.

Sometimes there are two common isotopes and this can have a greater effect on the relative atomic mass. About 70% of copper atoms have a mass number of 63, and the remaining 30% have a mass number of 65. This results in an average value of 63.6 for the relative atomic mass. Another example is selenium: 50% of selenium atoms have a mass number of 80, and the remaining isotopes have mass numbers of 76, 77 and 78. As a result, the relative atomic mass of selenium is 79, even though there is no common isotope of selenium with a mass number of 79.

To calculate the relative atomic mass of an isotope, you must take a weighted average of the masses of the isotopes. To do this you need the equation:

$$\text{Relative atomic mass} = (\text{fraction of isotope 1} \times \text{mass of isotope 1}) + (\text{fraction of isotope 2} \times \text{mass of isotope 2})$$

The fraction is the same as the percentage divided by 100. For example, a percentage of 50% is equal to a fraction of 0.5.

Let's look again at the example of copper. 70% of copper atoms have a mass number of 63 and the rest have a mass number of 65. We can calculate the relative atomic mass using the equation above.

$$\text{Relative atomic mass of copper} = 0.7 \times 63 + 0.3 \times 65 = 44.1 + 19.5 = 63.6$$

3. Calculate the relative atomic mass of chlorine. 76% of chlorine atoms have a mass number of 35. The remaining 24% have a mass number of 37.

4. Use the table below to answer the questions that follow.

Atom	Atomic number	Mass number
A	6	13
B	7	13
C	7	14
D	8	16

a) Which atom is carbon?

b) Which two atoms are isotopes?

c) Which atom has the fewest neutrons?

What types of atoms are there?

Dalton's original atomic theory gave five elements: hydrogen, carbon, nitrogen, oxygen and sulfur. Later revisions of his work increased the number of elements to 20, and then to 36. Throughout the 19th century, many more elements were discovered.

Some of these elements have very many protons in the nucleus. Elements usually have at least the same number of neutrons, and usually more. This means that the nucleus will be very big if there is a high number of protons. The nucleus will still be very small compared to the size of the atom however.

When a nucleus contains more than 82 protons, it becomes unstable as the forces inside the nucleus struggle to hold it together. Such nuclei are radioactive and they can break down, or decay, into smaller nuclei. The higher the number of protons, the more unstable the nucleus, and nuclear decay occurs more quickly. This makes heavier elements harder to observe.

Of all the elements, only 80 are stable. A further 14 radioactive elements such as uranium, thorium and plutonium exist naturally on Earth. Other heavier elements have been discovered by an experiment in which heavy elements are bombarded with neutrons. These experiments have succeeded in making elements up to atomic number 118.

Nuclei can also be unstable if they have too many or too few neutrons. For example, carbon-12 has six protons and six neutrons in its nucleus, and is stable. Carbon-14, which is a naturally occurring isotope of carbon with two extra neutrons, is unstable. This instability of carbon-14 is an important part of carbon dating. Carbon dating is used to determine the age of objects that were once living, such as wood and bone. By analyzing the isotopes of carbon present in a sample, scientists can determine how much of the carbon-14 has decayed and use this to calculate how old the sample is.

▼ Analysis of the amount of carbon-14, an unstable isotope of carbon, can be used to determine the age of objects. Carbon dating was used to approximate the date of this detail from the Egyptian Book of the Dead (around 1070 BC)

Heavy elements

The table below shows some heavy elements that have been discovered over the past century.

Atomic number	Element	Year of discovery
95	americium	1944
100	fermium	1953
105	dubnium	1970
110	darmstadtium	1994
115	moscovium	2010
120	?	*yet to be discovered*
125	?	

1. On a copy of the axes, sketch a graph of the atomic number against the year of discovery.
2. Add a line of best fit to your graph.
3. If the trend were to continue, when would you expect the elements with atomic numbers 120 and 125 to be discovered?
4. How reliable do you think your prediction is?

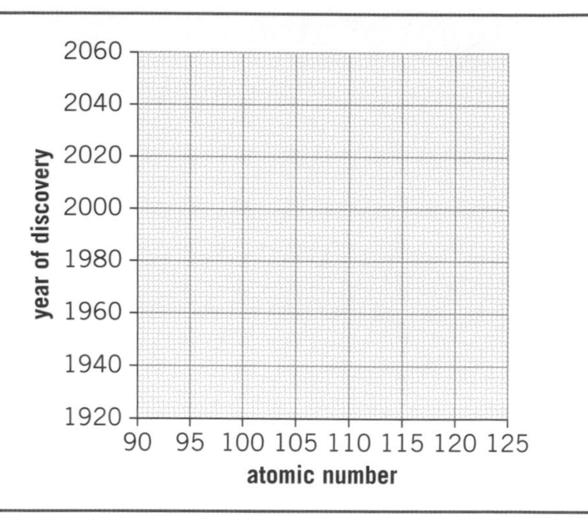

How can we organize the elements?

As individuals, we all have characteristics that define who we are. You might share some of these characteristics with your classmates—for example, there might be a group of you who all have brown eyes. Some atoms also have similar characteristics, or properties. Atoms can therefore be arranged into groups that have similar properties.

Metals are an example of elements having similar properties. They tend to be shiny, bendable and able to conduct electricity (see Chapter 8, Properties of matter, for more on the properties of metals). As a result, elements are often defined as metals or non-metals, although there are a few elements, such as silicon, which have properties that are between a metal and a non-metal. Such elements are called metalloids.

Some elements have more specific similarities. For example, lithium, sodium and potassium are all highly reactive metals that are soft and not very dense. In the 1860s, it was noted that these elements appeared to follow a pattern: the atomic masses of these elements increased by a regular amount. It led Dmitri Mendeleev, a Russian chemist, to publish an arrangement of the elements in 1869 that he called the periodic table.

In order to make the pattern work, Mendeleev sometimes had to leave gaps for elements that had not yet been discovered. Elements such as germanium and gallium were discovered later that fitted these gaps, confirming Mendeleev's arrangement of the elements. Some of the elements did not fit well and Mendeleev concluded that their atomic masses might be incorrect.

Today, the periodic table is widely used by chemists as the definitive arrangement of the elements.

							1 H hydrogen 1										2 He helium 4
3 Li lithium 7	4 Be beryllium 9											5 B boron 11	6 C carbon 12	7 N nitrogen 14	8 O oxygen 16	9 F fluorine 19	10 Ne neon 20
11 Na sodium 23	12 Mg magnesium 24											13 Al aluminium 27	14 Si silicon 28	15 P phosphorus 31	16 S sulfur 32	17 Cl chlorine 35.5	18 Ar argon 40
19 K potassium 39	20 Ca calcium 40	21 Sc scandium 45	22 Ti titanium 48	23 V vanadium 51	24 Cr chromium 52	25 Mn manganese 55	26 Fe iron 56	27 Co cobalt 59	28 Ni nickel 59	29 Cu copper 64	30 Zn zinc 65	31 Ga gallium 70	32 Ge germanium 73	33 As arsenic 75	34 Se selenium 79	35 Br bromine 80	36 Kr krypton 84
37 Rb rubidium 85	38 Sr strontium 88	39 Y yttrium 89	40 Zr zirconium 91	41 Nb niobium 93	42 Mo molybdenum 96	43 Tc technetium –	44 Ru ruthenium 101	45 Rh rhodium 103	46 Pd palladium 106	47 Ag silver 108	48 Cd cadmium 112	49 In indium 115	50 Sn tin 117	51 Sb antimony 122	52 Te tellurium 128	53 I iodine 127	54 Xe xenon 131
55 Cs caesium 133	56 Ba barium 137	57–71 lanthanoids	72 Hf hafnium 178	73 Ta tantalum 181	74 W tungsten 184	75 Re rhenium 186	76 Os osmium 190	77 Ir iridium 192	78 Pt platinum 195	79 Au gold 197	80 Hg mercury 201	81 Tl thallium 204	82 Pb lead 207	83 Bi bismuth 209	84 Po polonium –	85 At astatine –	86 Rn radon –
87 Fr francium –	88 Ra radium –	89–103 actinoids	104 Rf rutherfordium –	105 Db dubnium –	106 Sg seaborgium –	107 Bh bohrium –	108 Hs hassium –	109 Mt meitnerium –	110 Ds darmstadtium –	111 Rg roentgenium –	112 Cn copernicium –	113 Nh nihonium –	114 Fl flerovium –	115 Mc moscovium –	116 Lv livermorium –	117 Ts tennessine –	118 Og oganesson –

lanthanoids	57 La lanthanum 139	58 Ce cerium 140	59 Pr praseodymium 141	60 Nd neodymium 144	61 Pm promethium –	62 Sm samarium 150	63 Eu europium 152	64 Gd gadolinium 157	65 Tb terbium 159	66 Dy dysprosium 163	67 Ho holmium 165	68 Er erbium 167	69 Tm thulium 169	70 Yb ytterbium 173	71 Lu lutetium 175
actinoids	89 Ac actinium –	90 Th thorium 232	91 Pa protactinium 231	92 U uranium 238	93 Np neptunium –	94 Pu plutonium –	95 Am americium –	96 Cm curium –	97 Bk berkelium –	98 Cf californium –	99 Es einsteinium –	100 Fm fermium –	101 Md mendelevium –	102 No nobelium –	103 Lr lawrencium –

▲ The periodic table is the standard arrangement of the elements. The top number in each box gives the atomic number and the bottom number is the relative atomic mass

ATL **Communication skills**

Organizing and depicting information logically

Scientists often have to organize information in a way that is logical. The periodic table is a good example of this.

1. In which ways is the periodic table a better way of organizing the information than a list? Are there any benefits to a list?
2. Are there other ways in which the information could be presented?
3. Use an internet search to find a picture of the periodic table depicted as a spiral. Is this a good depiction of the information?

How can we combine atoms?

There are only 94 naturally occurring elements and of these, about 60 are found in the human body, although we only know the purpose of about 30. However, these elements can join together in chemical reactions to form millions of different chemical **compounds**. Some of the compounds are simple, such as water: a water **molecule** contains two hydrogen atoms and one oxygen atom. Others, such as DNA, contain more than one hundred billion atoms.

ABC

A **mixture** is two or more substances mixed together, but without changing the individual molecules of the original substances.

A **molecule** is two or more atoms chemically bonded together.

A **compound** is a molecule consisting of more than one type of atom.

The hydrogen and oxygen atoms that form water are not just mixed together. A **mixture** of two substances would contain the individual atoms or molecules of each and the properties would be similar to the original substances. For example, hydrogen and oxygen by themselves both exist as a gas—mixing them together would still give a gas.

To make a compound of water, the hydrogen and oxygen atoms have to be chemically bonded together, with the two hydrogen atoms bonded to a single oxygen atom. Each of these bonded units is called a molecule.

The hydrogen and oxygen atoms also form molecules when they are individual gases. Hydrogen atoms bond in pairs to form hydrogen molecules, and oxygen atoms also bond in pairs to form oxygen molecules. Because each hydrogen or oxygen molecule only contains one type of atom, we refer to it as an element. Water contains two types of atom and this is called a compound. Any molecule with more than one type of atom is a compound.

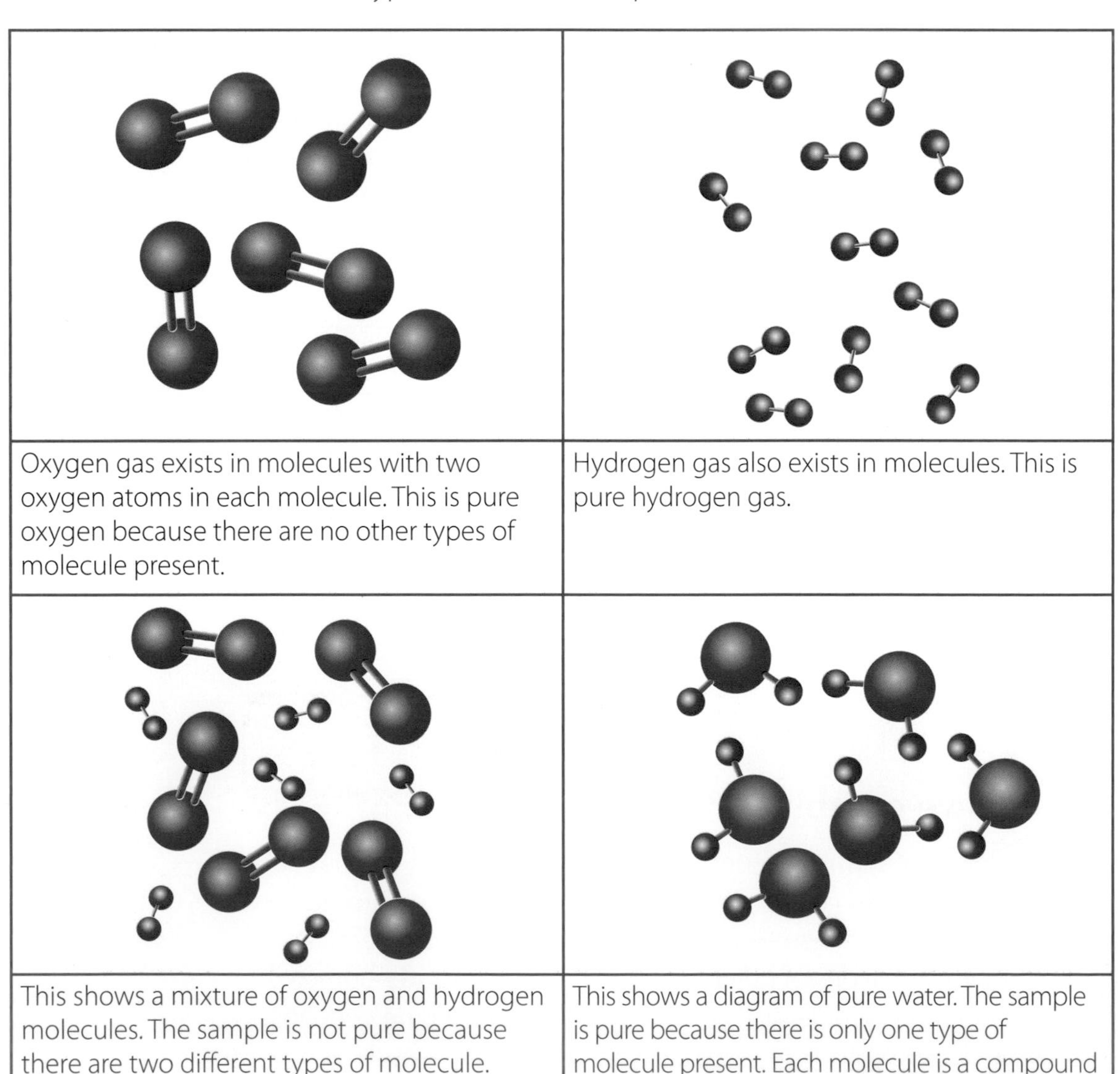

Oxygen gas exists in molecules with two oxygen atoms in each molecule. This is pure oxygen because there are no other types of molecule present.

Hydrogen gas also exists in molecules. This is pure hydrogen gas.

This shows a mixture of oxygen and hydrogen molecules. The sample is not pure because there are two different types of molecule.

This shows a diagram of pure water. The sample is pure because there is only one type of molecule present. Each molecule is a compound because it has more than one type of element.

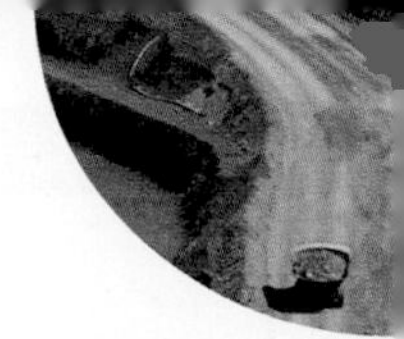

How can chemicals mix?

Chemicals can be mixed together in a variety of ways. Gases mix together easily as they can blend together. The nature of solids means that it is hard for them to mix together unless they are blended as a powder. Some liquids mix together well, and some do not, such as oil and water. Liquids that do not mix together are called **immiscible**, whereas those that do mix are called **miscible**.

A common way in which chemicals may form a mixture is as a **solution**. In a solution, a liquid, called the **solvent**, absorbs another substance, called the **solute**. For example, sugar dissolves in water. The solvent in this case is the water and the solute is the sugar. Often solutes are solids, but they can also be other liquids or gases.

The most common solvent is water. Water is able to dissolve many substances and life on Earth has evolved to make use of this. All life uses water in some way. In humans, sugars dissolve in the water in our blood to enable it to carry energy to our cells. Waste substances also dissolve in our blood and are removed. Water lines our lungs to enable oxygen to dissolve and be absorbed into our bloodstream.

Sometimes, solids and liquids can mix together such that the solid is divided into very fine lumps that spread out across the liquid. Muddy water is a good example of this. Although this might look like a solution, there are two clues that indicate that it is not: first, if the solvent is clear then the solution will also be clear; second, if the muddy water is left in a beaker for a while, the mud will settle to the bottom. This sort of mixture is called a **suspension**.

▲ The Yellow River, or Huang He, is the second longest river in China. Its name comes from the distinctive color of the water, which is caused by the mud that it carries downstream. This mixture of mud and water is a suspension. As the river gets closer to the sea, it gets wider and flows slower. When this happens, the mud comes out of suspension and settles to the bottom

▲ Seawater is a solution. Each kilogram of seawater contains about 35 g of sodium chloride dissolved in it. Other elements such as potassium and bromine are also found dissolved in seawater. Gases can also dissolve in water—oxygen dissolves in water, which enables fish to survive

▲ When oil is poured into water, it forms a layer on top. This is because oil and water are immiscible liquids

ABC

Two **miscible** liquids are able to completely dissolve in each other.

Immiscible liquids do not mix and will separate into layers.

A **solution** is a mixture of substances where a liquid is able to absorb another substance.

A **solute** is the substance that is dissolved in a solution.

A **solvent** is the liquid in which a solution is formed.

A **soluble** substance is one that can be dissolved.

A **suspension** is a mixture of fine solid particles in a liquid.

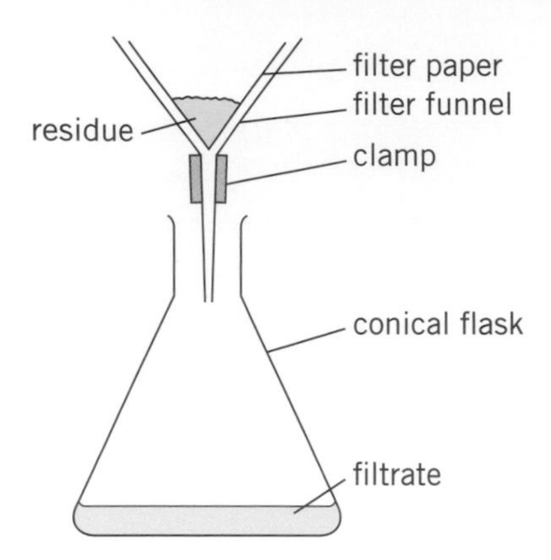

▲ This is the apparatus used for filtration. Filtration is a process that separates a solid that is suspended in a liquid

The **residue** is the solid material that is separated out in filtration or evaporation.

The **filtrate** is the liquid that passes through the filter paper.

How can we separate chemicals?

Mixtures of chemicals can be separated using physical techniques. The word physical refers to changes that do not involve any chemical reactions (see Chapter 9, Chemical reactions, for what a chemical reaction is). Physical changes can include dissolving, evaporating or condensing a substance (see Chapter 8, Properties of matter, for more on how materials can change state such as evaporating or condensing). Some important methods for separating mixtures are **filtration**, **evaporation** and **distillation**.

Filtration can be used to separate a suspension of a liquid and a solid. In this method, filter paper is used to separate solids and liquids—filter paper contains tiny holes that liquid can pass through, but solid substances that are in suspension will not fit through the holes and remain on top of the paper. The solid that is removed from the mixture is called the **residue** and the liquid is called the **filtrate**.

This method can also useful for separating two solids where only one can be dissolved in a certain solvent. In this case, the two solids are added to the solvent. The **soluble** solid dissolves in the solvent and passes through the filter paper, and the insoluble one is filtered out. The soluble solid can be recovered by evaporating the solvent away. The insoluble solid would need to be washed with pure solvent before it is dried.

Experiment

Separating salt and sand

Your teacher will provide a mixture of salt (sodium chloride) and sand.

Method

- Take about 5 g of the mixture and place it in a beaker.
- Add 50 mL of distilled water to your beaker and stir to dissolve the salt.
- Place filter paper into a funnel and clamp this above the neck of a conical flask. Filter the salt–sand mixture.
- Pour the filtered solution into an evaporating basin and heat gently to evaporate the water. Once you see solid salt start to form, stop heating the basin and leave the basin to allow the remainder of the water to evaporate.
- Wash the filter paper and sand with another 50 mL of distilled water.
- Use a spatula to remove the sand from the filter paper and place on a new dry piece of filter paper.

Questions

1. Why was distilled water used in this experiment?
2. Why was a second amount of distilled water poured onto the sand?

Evaporation is useful separation technique when the mixture consists of a solid substance that is dissolved in a solvent. Gently heating the solution will evaporate the solvent, leaving the solid as a residue. Evaporation is only useful when it is the solid, rather than the solvent, that you wish to obtain. If you wanted the solvent from the mixture, then you would use distillation.

Distillation is similar to evaporation: the mixture is heated so that the solvent evaporates. However, the solvent vapor is then passed through a condenser so that it can then be collected. Fractional distillation is a method of distillation that can be used to separate a mixture of two liquids. If the liquids have different boiling points, then the mixture can be heated so that one liquid evaporates, leaving the other behind. Fractional distillation does not completely separate the mixture: some of the liquid with the higher boiling point will evaporate, while some of the liquid with the lower boiling point will remain dissolved.

ABC

Distillation is a separation technique where a liquid is boiled and the vapor is condensed to obtain a purer sample of the liquid.

Evaporation is a separation technique where a solution is heated to leave behind a solid residue.

Filtration is a separation technique where a liquid is passed through a filter to remove any solids.

◀ Distillation takes place on a large scale in these towers. Crude oil is distilled to separate different substances. These provide fuels such as gasoline and diesel

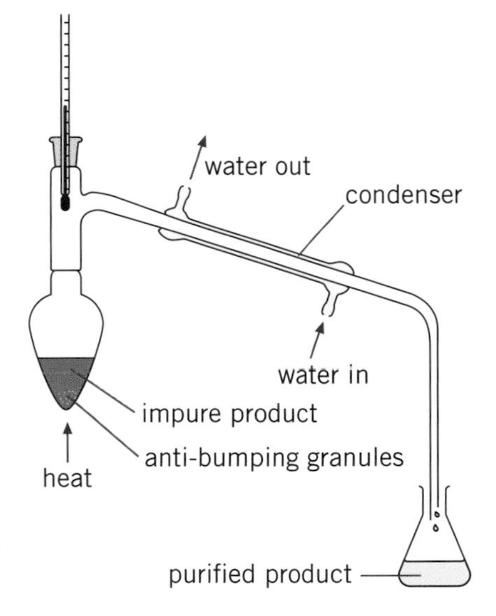

▶ This is the apparatus that is used for distillation. This process is used to separate two miscible liquids. The liquid that has the lower boiling point will evaporate more easily. It will pass through the condenser where it returns back to a liquid and collects in the flask

1. Describe the processes that you would use to separate:
   a) the soil from some muddy water
   b) the copper carbonate from a mixture of copper carbonate and copper sulfate. (Hint: copper sulfate is soluble in water and copper carbonate is insoluble in water.)

How do we name simple chemical compounds?

Some chemical compounds, particularly those that occur in living organisms, can be very complicated and consist of huge numbers of atoms. A single molecule of DNA, the molecule that defines our genetic identity, can have over 200 billion atoms.

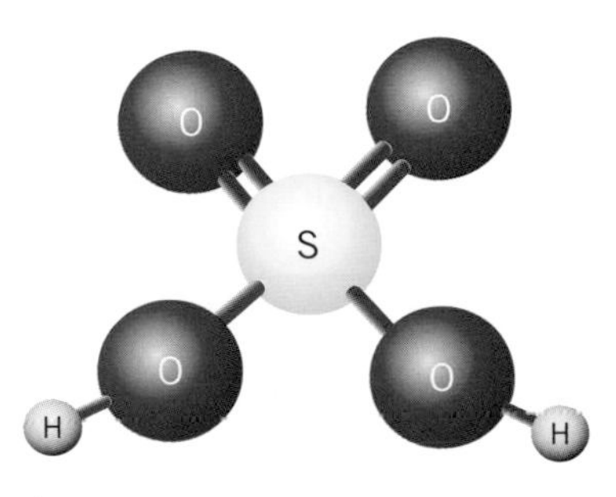

▲ This is a diagram of a molecule of sulfuric acid. Its chemical formula is H_2SO_4

Naming and describing these compounds can be very difficult. However, there are some easier ways in which scientists describe simple chemical compounds that contain only a few atoms.

Chemical formula is a notation using chemical symbols to describe the atoms that make up a molecule.

A water molecule is an example of a simple chemical compound; each molecule consists of two hydrogen atoms and one oxygen atom. As a result, the structure of the molecule can be written as H_2O. H and O are the chemical symbols for the elements hydrogen and oxygen, and the small 2 after the H indicates that there are two hydrogen atoms. This is called its **chemical formula**. In a similar way, the molecule of hydrogen consisting of two hydrogen atoms has a chemical formula H_2. As described in the experiment below, iron sulfide has the chemical formula FeS. The periodic table is a useful place to find the names of chemical elements and their chemical symbols.

Experiment

Investigating mixtures and compounds

Safety

- Wear safety glasses at all times.
- Powdered iron is a flammable solid. Store the container away from the open flame.
- Powdered sulfur causes skin irritation so gloves must be worn at all times.

The compound iron sulfide has the chemical formula FeS. This means an iron sulfide molecule has one iron atom and one sulfur atom bonded together.

Method

- Your teacher will prepare a mixture of powdered iron and powdered sulfur. (This should be in a 7:4 ratio of iron to sulfur).
- Weigh approximately 1 g of the mixture and put it in a test tube.
- Place a piece of mineral wool in the mouth of the test tube.
- Carefully heat the iron and sulfur mixture with a Bunsen burner. When it starts to glow, remove the test tube from the flame and turn the Bunsen burner off.
- Leave the test tube to cool.
- When the test tube has cooled, examine the product of the reaction. How does it differ from the initial mixture?
- Test the products of the reaction with a magnet. The initial iron was magnetic—is the iron sulfide magnetic?

Questions

1. How would you separate the original mixture of iron and sulfur?
2. What evidence is there that iron sulfide is a different chemical compound to the iron and sulfur that you started with?
3. A molecule of iron sulfide has one iron atom and one sulfur atom. Why do you think the original mixture of iron and sulfur had a greater mass of iron?

Compounds can also be described using words. For example, the gas CO_2 is called carbon dioxide. The name is determined by the atoms in the molecule. The first element that appears in the name is the one that appears the furthest to the bottom left of the periodic table. If there are only two types of atom in the compound, the name of the second element has its ending replaced with the suffix -ide. Hence oxygen becomes oxide and chlorine becomes chloride.

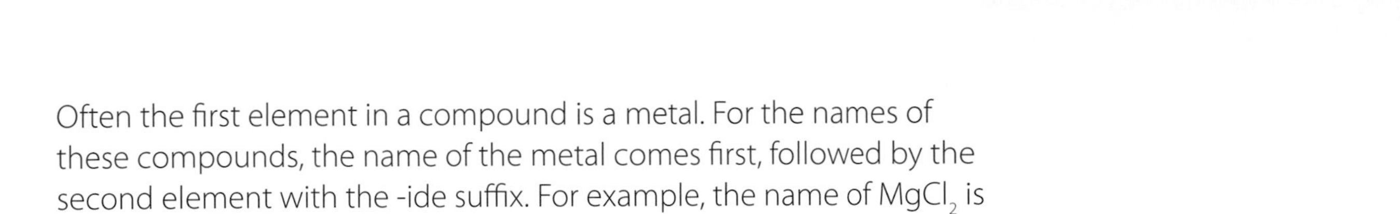

Often the first element in a compound is a metal. For the names of these compounds, the name of the metal comes first, followed by the second element with the -ide suffix. For example, the name of $MgCl_2$ is magnesium chloride.

However, if the first element is a non-metal, the number of atoms of each type needs to be indicated with a prefix. For example, as there are two oxygen atoms in CO_2, the prefix di- appears before oxide in the name of the compound. The overall name of the compound is carbon dioxide.

If there is only one atom of the first element in the compound, no prefix is used. This is why CO_2 is not called monocarbon dioxide. However, the mono- prefix is used for the second element. As a result, the gas CO is named carbon monoxide. The prefixes that are used are shown in the table below.

Number of atoms	Prefix	Example	Chemical formula
1	mono-	carbon monoxide	CO
2	di-	carbon dioxide	CO_2
3	tri-	sulfur trioxide	SO_3
4	tetra-	silicon tetrafluoride	SiF_4
5	penta-	phosphorus pentachloride	PCl_5
6	hexa-	sulfur hexafluoride	SF_6

Sometimes, compounds contain three types of atom, one of which is oxygen. In this case, the suffix -ide is replaced with –ate, which represents the presence of oxygen. For example, sodium silicate contains three types of atoms: sodium, silicon and oxygen.

1. Identify the chemical formula and the name of the compounds shown below. You may wish to use a periodic table to identify the elements from their chemical symbols.

a) **b)**

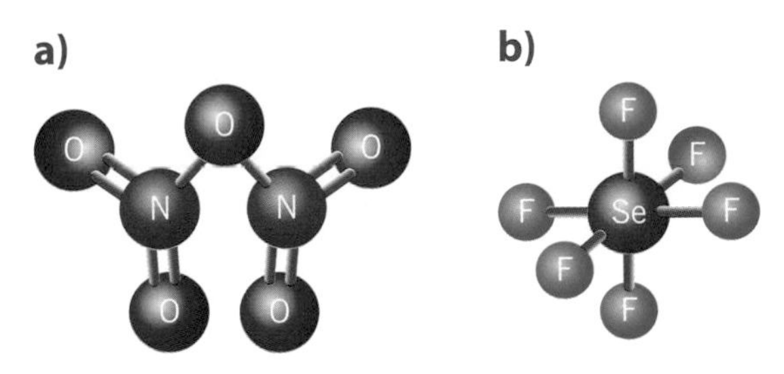

2. Identify the elements present in the following compounds:
   a) lithium nitride
   b) iron sulfate
   c) calcium carbonate.

How can we find the mass of a molecule?

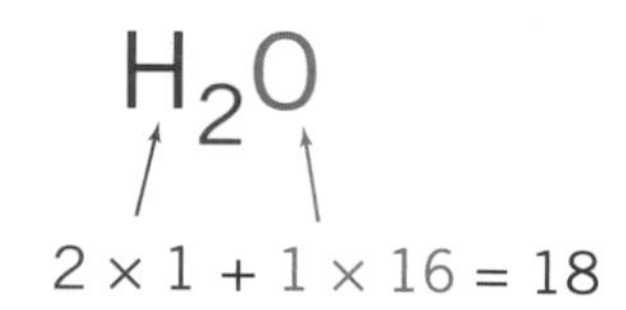

▲ The relative molecular mass of water is 18

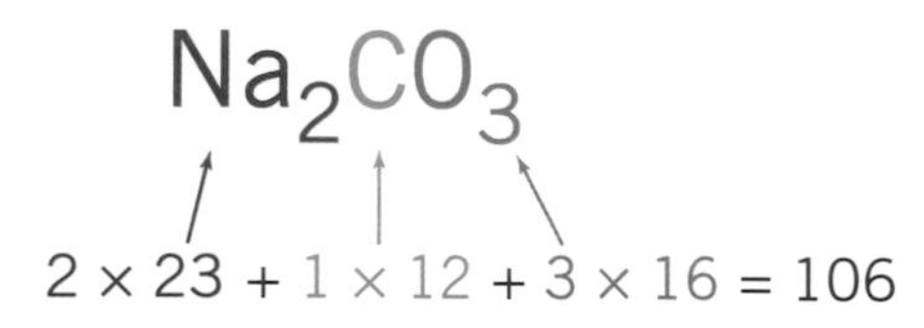

▲ The relative molecular mass of sodium carbonate is 106

Just as different atoms have a relative atomic mass, molecules also have a relative mass. This is called the relative molecular mass or relative formula mass.

To find the relative molecular mass, look at the chemical formula of the molecule. Let's use H_2O as an example. The small subscript 2 shows that there are two hydrogen atoms in the molecule. The relative molecular mass is the sum of the relative atomic masses of each atom. In H_2O, there are two hydrogen atoms and one oxygen atom. The periodic table shows that hydrogen has a relative atomic mass of 1, and oxygen has a relative atomic mass of 16. As a result, the relative molecular mass of H_2O is $2 \times 1 + 16 = 18$.

The relative molecular mass of more complicated compounds can be found in a similar way. For example, the relative molecular mass of sodium carbonate, Na_2CO_3 is 106.

Finding the relative molecular mass of DNA bases

The chemical responsible for our identities is DNA, as it carries our genetic identity. Although it is a very complicated molecule, it consists of a sequence of just four different base units. The four base units of DNA are called adenine, guanine, cytosine and thymine. Adenine has the chemical formula $C_{10}H_{12}O_5N_5P$, guanine has the formula $C_{10}H_{12}O_6N_5P$, cytosine is $C_9H_{12}O_6N_3P$ and thymine is $C_{10}H_{13}O_7N_2P$.

The table below gives the masses of each type of atom.

Element	Chemical symbol	Relative atomic mass
hydrogen	H	1
carbon	C	12
nitrogen	N	14
oxygen	O	16
phosphorus	P	40

1. Calculate the relative molecular masses of each of the DNA base units.
2. In a DNA molecule, base units pair together. The base units in a pair have a similar size in order to maintain the structure of the DNA molecule. Determine the way in which the four bases pair together such that the bases in a pair have a similar relative molecular mass.

1. Identify the name and find the relative formula mass of the following compounds. You may wish to use the periodic table on page 129 to help you.
   a) NaCl
   b) $CaBr_2$
   c) SiO_2
   d) SF_4
   e) $MgCO_3$

Summative assessment

Statement of inquiry:

The complex chemicals that enable life to exist are formed from only a few different types of atom.

This assessment is based on the environmental effects that atoms and molecules can have, and the importance of considering sustainability when thinking about chemicals and their uses.

Atoms, elements and compounds in the environment

1. The atmosphere mainly consists of nitrogen and oxygen. Which term describes the combination of gases in the atmosphere?

   A. Compound **C.** Mixture
   B. Element **D.** Solution

2. Ozone is an important constituent of the stratosphere as it blocks harmful UV rays from the Sun. Ozone has the chemical formula O_3. Which word describes ozone?

   A. Atom **C.** Element
   B. Compound **D.** Isotope

3. Chlorofluorocarbons (or CFCs) are chemicals that destroy ozone in the atmosphere. A common CFC has the chemical formula CCl_2F_2. Using the table below, what is the relative formula mass of CCl_2F_2?

Chemical symbol	Element	Relative atomic mass
C	carbon	12
Cl	chlorine	35.5
F	fluorine	19

   A. 60.5
   B. 66.5
   C. 121
   D. 133

4. Calcium nitrate and potassium sulfate are both used as fertilizers. If too much of them are used, then they can pollute the environment. Which statement about these two substances is true?

   A. Both are impure mixtures.
   B. Both molecules consist of two types of atom.
   C. Both molecules contain oxygen atoms.
   D. Both are pure elements.

5. The testing of nuclear weapons in the 20th century caused the amount of caesium-137 to increase. Unlike caesium-133, which is stable, caesium-137 is unstable and radioactive. Which statement is true?

   A. Caesium-137 is a different element to caesium-133.
   B. Caesium-137 has more protons that caesium-133.
   C. Caesium-137 has the same relative atomic mass as caesium-133.
   D. Caesium-137 has more neutrons than caesium-133.

6. Water that flows through rivers may have compounds dissolved in it that make it unsafe to drink. Which technique could be used to remove the dissolved substances and obtain the pure water?

   A. Boiling
   B. Distillation
   C. Evaporation
   D. Filtration

Investigating air pollution

7. A pupil is investigating air pollution. She acquires a portable gas monitor that can measure the concentration of pollutant gases in the air. She decides to use the gas monitor to measure air pollution in five different places:
   - Inside her home
   - At school
   - By the nearest busy road
   - In the center of her nearest town
   - In the countryside.

   a) Identify the independent and dependent variables for this experiment. [2]

   b) Suggest how she might present her results. [1]

   c) She decides that her experiment would be improved by measuring the air in the morning, the afternoon and the evening and taking an average of her values. Explain why this is an improvement to her method. [2]

   d) She now tries to decide whether to take all the readings for all the different places on one day, or to measure each place on a different day. Discuss the advantages and disadvantages of each method. [3]

   She decides to test the gas monitor's sensitivity to nitrogen dioxide, which is a harmful polluting gas. She finds out that nitrogen dioxide is released when copper nitrate is heated. She plans to heat a test tube of copper nitrate over a Bunsen burner and see how much nitrogen dioxide the gas monitor detects.

   e) What safety precautions should she take if she conducts this experiment? [2]

▲ Air pollution can be very different in different locations

CFCs in the atmosphere

Trichlorofluoromethane (CCl_3F) is a CFC gas that was used in refrigerators. Its use was banned by the Montreal agreement in 1987, because of the damage that CFC gases do to the ozone layer.

8. The amount of CCl_3F that was detected in the atmosphere in different years is shown in the table on the left.

   a) Plot a graph of the data, with amount of CCl_3F on the *y*-axis and year on the *x*-axis. [4]

Year	Amount of CCl_3F (parts per trillion)
1980	153
1985	196
1990	245
1995	262
2000	258
2005	249
2010	240
2015	232

b) Use your graph to estimate the following:

i) the rate at which the amount of CCl_3F was increasing per year between 1980 and 1990. [2]

ii) the length of time between the Montreal agreement, which banned the use of CFCs in 1987, and the peak amount of CCl_3F in the atmosphere. [1]

iii) if levels of CCl_3F continue to fall in the same way, the year they will be below 200 parts per trillion. [3]

Atom economy

In 1998, Paul Anastas and John Warner wrote a book called *Green Chemistry: Theory and Practice*. It was published by Oxford University Press. In the book, they outlined 12 principles for green chemistry. One of these principles was atom economy.

9. Write a suitable reference for the book *Green Chemistry: Theory and Practice*. (See the ATL feature in Chapter 1, Motion, for how to give a reference.) [1]

Hydrogen is an important gas that is used for many chemical processes. It is often obtained by reacting methane (CH_4) with steam in the following reaction.

$$CH_4 + H_2O \rightarrow CO + 3H_2$$

10. Using the periodic table, calculate the relative formula mass of the methane and water. [2]

The atom economy of a reaction can be calculated using the formula:

$$\text{atom economy} = \frac{\text{sum of relative formula masses of the useful product(s)}}{\text{sum of relative formula masses of reactants}}$$

11. The useful product in this reaction is hydrogen gas. Hydrogen molecules have a relative formula mass of 2. The "3" before H_2 in the chemical equation indicates there are three hydrogen molecules. Calculate the sum of the relative formula masses of the useful product, and the sum of the relative formula masses of the reactants. [2]

12. Calculate the atom economy of this reaction. [2]

13. The concept of atom economy, and the other principles of green chemistry, helps to encourage industrial chemists to reduce waste. Suggest one benefit and one problem of reducing chemical waste from industrial processes. [2]

8 Properties of matter

Key concept: Relationships

Related concepts: Form, Transformation

Global context: Globalization and sustainability

We normally think of oxygen as being a gas. However, when it is cooled to −183°C, it condenses into a liquid. Liquid oxygen is pale blue and slightly magnetic. It is also very reactive—because oxygen is required for burning, objects that are soaked in liquid oxygen burn far more vigorously and even explosively. What else is required for burning to occur?

Plastic has many very useful properties. Although there are many types of plastics, they are generally low density, waterproof, cheap, durable and can be easily molded into shape. As a result, plastic has been an increasingly popular material for many products. What are the problems of using plastic?

Statement of inquiry:

New materials with differing properties have helped to create today's global society and may hold the answers to some of the problems of the future.

▲ Gymnasts train hard to be strong and flexible. Their muscles, joints and connective tissues need to be elastic but also tough. What other objects need to be designed to be strong but also flexible?

◄ Many spiders can produce different types of silk with different properties for different purposes. The threads that run from the center to the outside support the web and can be tougher than the same mass of steel. They are not sticky so the spider can walk along these without becoming tangled in its own web. The silk used for the threads that run around the web in a spiral pattern is sticky (to capture the spider's prey) and more stretchy. Which other organisms produce materials with useful physical properties?

Key concept: Relationships

Related concepts: Form, Transformation

Global context: Globalization and sustainability

Statement of inquiry:

New materials with differing properties have helped to create today's global society and may hold the answers to some of the problems of the future.

Introduction

Matter is made of atoms and there are fewer than 100 naturally occurring types of atom that make up the world around us (see Chapter 7, Atoms, elements and compounds). However, even though the world is made up of a small number of different types of building blocks, the matter around us can take on many different forms and have many different properties. For example, the air around us is a gas that we can easily move through. On the other hand, we cannot move through solid objects such as a wall. In this chapter, we will look at different states of matter and how they behave.

We shall see how matter is formed from atoms and molecules and how they are arranged. For this reason, one of the related concepts of the chapter is form. We will also see how matter can change between different states, and so the other related concept is transformation.

Even matter in the same state can have very different properties. For example, solids may be brittle and fragile like glass, tough and bendable like metals or soft and easily melted like wax. The way materials behave is related to the behavior of their atoms and molecules. The different way in which they interact causes properties such as boiling points and melting points to differ. There is also a relationship between the properties of different materials and the way we use them. As a result, the key concept of this chapter is relationships.

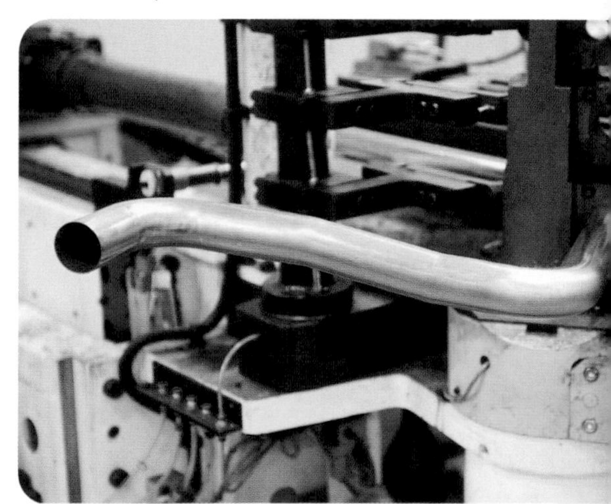

▲ Although they are all solids, these materials all behave differently. Wax is soft and easily melted, glass is brittle and fragile, while metal is tough and can be bent

Understanding how matter behaves has been fundamental ever since humans first started to make things. Picking the materials with the most useful properties has helped us to build the civilization around us. The conductivity of metals enabled us to connect electricity to homes and use wires to build communications between cities and countries. Today, scientists try to develop new materials that will allow our civilization to grow in a more sustainable way. The global context of the chapter is globalization and sustainability.

The electrical properties of silicon are unusual in that it does not conduct electricity in the way that metals do, nor does it insulate like most non-metals. It is a semiconductor and its electrical properties have been used in the development of complicated electronic circuits. These circuits are fitted onto flat pieces of silicon called "chips", as in the picture above. This has enabled the technological advance of computing and electronics

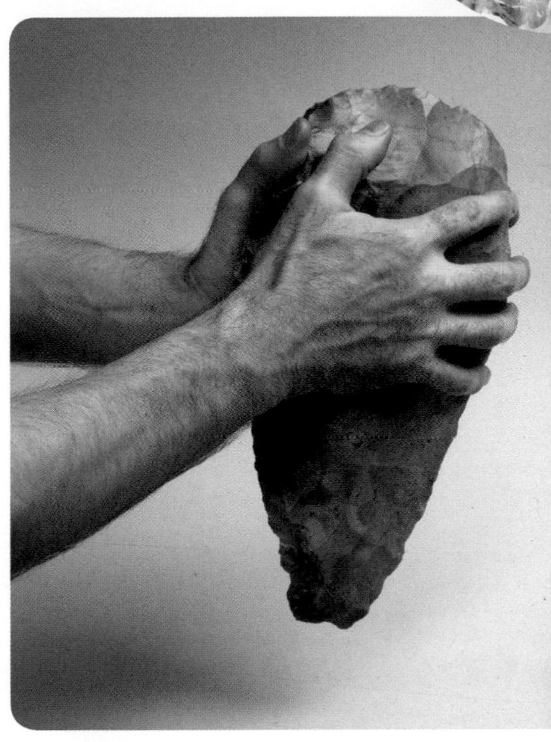

In the Paleolithic period, early human societies used stone to make tools. This hand axe dates from 400,000 to 300,000 BC. Today, engineers have a huge variety of materials with which to make tools, construct buildings and make useful devices

What are the states of matter?

The matter around us tends to fall into one of three different categories called **states of matter**, or **phases**. The three common states of matter are **solids**, **liquids** and **gases**.

Solids have a surface and a fixed shape. The is because the molecules that make up a solid are packed very closely together. As a result, the forces between the molecules are very strong and the molecules are fixed in place. Because the molecules are so close together, it is very hard to push them any closer. Therefore solids are not compressible. Equally, it is hard to pull the molecules further apart and so this gives solids their fixed volume.

Liquids are also hard to compress and expand and so have a fixed volume and a definite surface. However, unlike solids, they do not have a fixed shape. Instead, they can flow and will fill the container they are in, from the bottom upwards.

The reason for liquids behaving in this way is that their molecules are still closely packed. This means that liquids maintain a fixed volume because it is difficult to separate the molecules or push them closer together. However, the molecules are not in a fixed position and are able to move past each other. This gives liquids the ability to flow and change their shape according to the shape of their container.

Gases are the third state of matter. They can flow like liquids but do not have a fixed volume and have no definite surface. They take the shape of their container and expand to fill it. They are easy to compress and expand.

ABC A **state of matter**, or **phase**, is one of the three common forms in which materials exist: **solid**, **liquid** or **gas**.

The reason for a gas's behavior is that the molecules are far apart from each other. For example, the molecules in the air around us are about ten times further apart than molecules of water. As a result, the forces between the molecules are much weaker. It is therefore easy to move the molecules closer or further apart. This means that gases can expand and be compressed.

▲ The arrangements of the atoms and molecules determines the properties of solids, liquids and gases. The steel girders are solid because the atoms are tightly packed in fixed places. The water is a liquid and can flow through pipes because the atoms are able to move past each other. The gas inside the bubbles can expand and fill the bubble because the atoms are separated

▶ Many machines such as diggers and tractors use hydraulic pistons to control their movement and to provide large forces. These hydraulic pistons use the fact that liquids are incompressible to control the forces

How does the motion of molecules affect a material?

Heating a substance gives more energy to its particles. This makes them move faster. In a gas, molecules push more against the walls of their container. In solids, the molecules push against each other. In liquids, both of these things happen. The consequence is that heating a material causes it to expand. This is called **thermal expansion**.

Thermal expansion is when materials expand when they are heated. Most materials have this property.

Observing the thermal expansion and contraction of gases

Take two beakers. Place one in a fridge or freezer for ten minutes. Place the other in a warm place for the same length of time or stand it in hot water.

After ten minutes, cover each beaker with plastic food wrap so that they are sealed.

Place the cold beaker in a warm place or stand it in hot water. As the air inside the beaker warms up, the plastic wrap is stretched outwards as the air expands.

Place the warm beaker in the fridge or freezer. As the air cools, it contracts, and the plastic wrap is sucked inwards.

This picture shows a gap in a road that goes over a bridge. In hot weather, the road expands. The gap allows either side of the road to move together, rather than causing forces to act on the road and possibly damage it

How can molecules be arranged in a solid?

The arrangement of the individual molecules in a solid can only be seen on a tiny scale. However, their arrangement can have observable effects on the solid.

Some solids have a regular arrangement of their molecules that extends over large sizes. This highly regular arrangement causes them to form in regular shapes, or they can be easily cut into these shapes. Such solids are called crystals.

ABC A **crystal** is a solid where the atoms or molecules are ordered in a regular arrangement. As a result, they often form regular geometric shapes.

Experiment

Growing crystals

Sodium chloride crystals and copper sulfate crystals are different shapes because of the different arrangements of the atoms. You can grow crystals using the following method.

Safety

- Wear safety glasses at all times.
- Copper sulfate causes skin irritation so gloves must be worn at all times.
- Copper sulfate is very toxic to aquatic life so dispose of this chemical appropriately.

Method

- Make a small amount of saturated solution. You can do this by dissolving salt or copper sulfate in 100 to 200 cm^3 of boiling water until no more will dissolve.
- Pour this solution into shallow bowl, taking care to leave any undissolved salt or copper sulfate behind. Leave for at least 24 hours.

Crystals will form at the bottom and sides of the bowl. As the water evaporates, they grow bigger. You can use one of these crystals to grow a larger crystal.

- When the crystals are big enough, select one and tie a piece of thread around it.
- Make another saturated solution as before. You can do this by heating the solution that you have and dissolving more salt or copper sulfate in it until no more dissolves.
- Suspend your crystal in the saturated solution using the thread and leave it undisturbed for a few days.

To grow a big crystal, then you will need to re-saturate your solution every couple of days. To do this, remove the crystal and heat the solution before dissolving more salt or copper sulfate in it. Leave your solution to cool before you put your growing crystal back in.

▲ A seed crystal can be used to grow a larger crystal by suspending it in a saturated solution

Questions

1. Why do you think you dissolved the original substance (sodium chloride or copper sulfate) in hot water?
2. If you have been able to grow more than one crystal, compare the shapes of the different crystals.

▶ Iron pyrite (also known as fool's gold) has atoms of iron and sulfur arranged in a cubic structure. This causes the crystals to take on a cubic shape

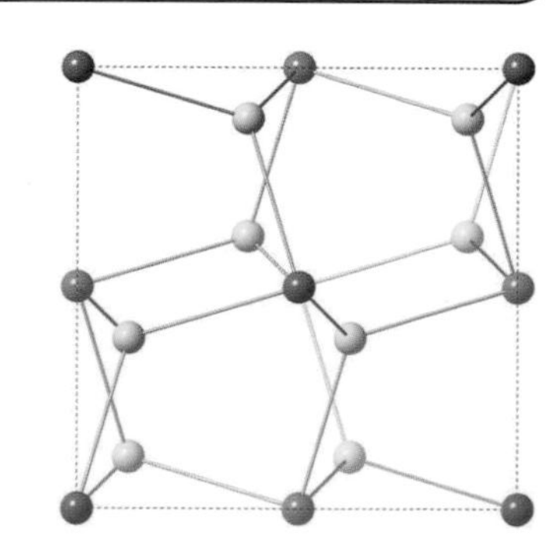

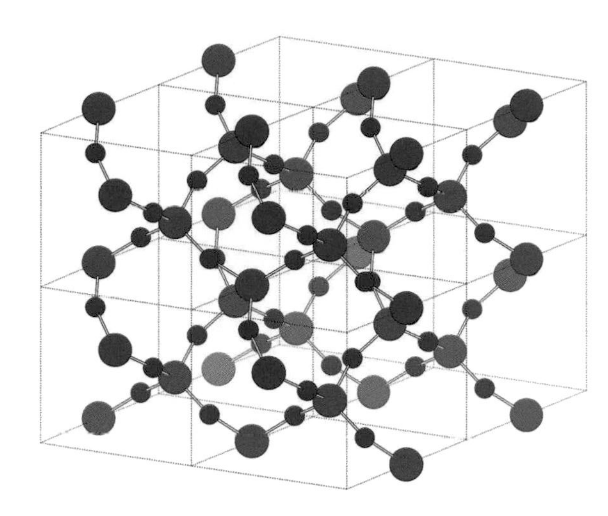

Quartz, on the other hand, has silicon and oxygen atoms arranged in a hexagonal pattern, which gives quartz crystals their hexagonal shape

Some solids have no long-range structure in the arrangement of their molecules. Such solids are described as being amorphous. Glass is an example of an amorphous solid. Because glass has no regular structure, when it breaks it shatters into irregular shapes

This turbine blade is designed to withstand the demands of an airplane's engine. In order to increase its strength, it is grown from a single crystal of iron mixed with other elements, such as tungsten

ABC

A **phase change** is a change in a material from one state of matter to another.

Melting is when a solid changes to a liquid state.

Freezing is when a liquid changes to a solid state.

Vaporization is when a liquid changes to a gaseous state.

Condensation is when a gas changes to a liquid state.

Sublimation is when a solid changes directly to a gaseous state.

The **melting point** is the temperature at which a solid changes to a liquid.

The **boiling point** is the temperature at which a liquid substance vaporizes into a gas.

How does matter change states?

A substance can exist in different states depending on its surroundings. For example, water is most commonly found as a liquid, but at cold temperatures it freezes and becomes a solid. If water is heated, it evaporates and becomes a gas. Changing a substance from one state of matter to another is called a **phase change**.

When a solid is heated, energy is given to the molecules. If they are given enough energy, they are able to break out of their fixed positions and start moving past each other. The solid has now become a liquid in a process called **melting**. The reverse process is called **freezing**, where molecules in a liquid lose energy and stop being able to move past each other.

If a liquid is heated, the molecules may be given enough energy to break apart from each other. The liquid has now become a gas in a process called **vaporization**. The reverse process is called **condensation**, when the molecules no longer have enough energy to break apart.

These phase changes usually occur at a particular temperature. For example, pure water freezes at 0°C and will boil at 100°C. The temperature at which a substance boils or condenses is called its **boiling point**, and the temperature at which a substance freezes or melts is called its **melting point**.

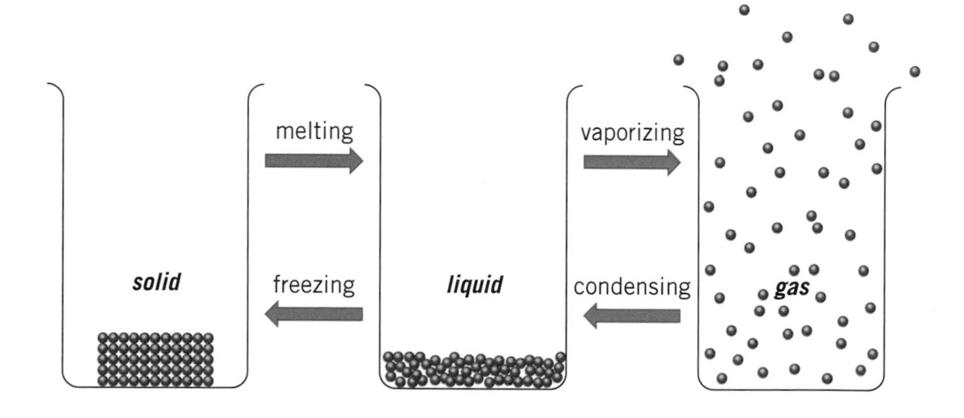

▶ When enough thermal energy is given to the molecules of a solid, they can break out of their fixed positions. The solid becomes a liquid in a process called melting. If the liquid is given more thermal energy, the molecules may be able to break apart from each other and the liquid becomes a gas in the process of vaporization. The reverse processes are condensing (gas to liquid) and freezing (liquid to solid)

Melting and boiling points are normally fixed temperatures for different substances, although they may change if a substance is at a different pressure. This is why water boils at a lower temperature at the top of a mountain where the pressure is lower. For example, water would boil at a temperature of 71°C at the top of Mount Everest. Whether a substance is pure or not will also affect the melting and boiling points.

For example, adding salt to water causes its melting point to be lower and its boiling point to be higher.

In cold weather, putting salt on roads lowers the temperature at which water freezes. This keeps the roads from becoming icy and dangerous

Matter can change state directly from a solid to a gas without being a liquid in between. This phase change is called **sublimation**. Solid carbon dioxide (often called dry ice) sublimes as it warms up to room temperature. Iodine also sublimes if it is heated.

Many other substances can sublime if they are at low pressures. If the pressure is about 170 times less than atmospheric pressure, water will sublime. In space, where the pressure is very low, many substances sublime and so liquids are rare in space.

Dry ice is carbon dioxide that has been cooled until it is frozen. As it warms up, it does not pass through the liquid state, but instead sublimes from a solid directly into a gas

Freezing point of salt water

In an experiment to investigate the freezing point of salt water, salt was added in varying quantities to 100 cm^3 of water. The water was then cooled to find its freezing point. The results of the experiment are shown in the table below.

1. Plot a graph of the data with the mass of salt on the *x*-axis and the freezing point on the *y*-axis.
2. Add a line of best fit to your graph.
3. What control variable was used in this experiment?
4. The south end of the Great Salt Lake in the USA has about 14 g of salt for every 100 cm^3 of water. Use your graph to estimate its freezing point.
5. The Mediterranean Sea has 19 g of salt for every 500 cm^3 of water. Use your graph to estimate the freezing point of the Mediterranean Sea.

Mass of salt (g)	Freezing point (°C)
0	0
5	−3
10	−6
15	−10
20	−14
25	−18
30	−20

ABC **Density** is a measurement of how much mass is contained in a given volume of a substance. It is found by dividing the mass of material by its volume.

▼ Although the pumpkins on the left-hand side of the balance have a greater volume than the weight on the right, they both have the same mass and so they balance. This is because the weight has a greater density than the pumpkins

How dense is matter?

The way in which the molecules of a material are packed together also affects the **density** of a material. Density is amount of mass contained in a given volume of material and can be calculated using the equation:

$$\text{density} = \frac{\text{mass}}{\text{volume}}$$

Density is a useful property of matter because most solids and liquids have a constant value for their density. This value depends on the substance or material. If the molecules of a substance are packed closer together, the density will be higher. As a result, solids and liquids, where the molecules are packed closely together, have higher densities than gases, where the molecules are spaced further apart.

To measure the density of an object, you need to find its mass and its volume. The mass of an object can usually be measured with a balance. The volume may be more difficult to measure.

The volume of a rectangular solid can be found by measuring its length, width, and height and then using the equation:

$$\text{volume} = \text{length} \times \text{width} \times \text{height}$$

The volume of an irregularly shaped solid may be measured using a displacement can or a measuring cylinder. To use a measuring cylinder, enough water must be put in the measuring cylinder to completely submerge the object that is being measured. First, the volume of the water is measured by reading the scale on the measuring cylinder. The object is then put into the measuring cylinder and the water level will rise. The new volume can now be measured, and the increase in measured volume is equal to the volume of the solid.

1. A plank of wood is 3 cm thick, 16 cm wide and 75 cm long.

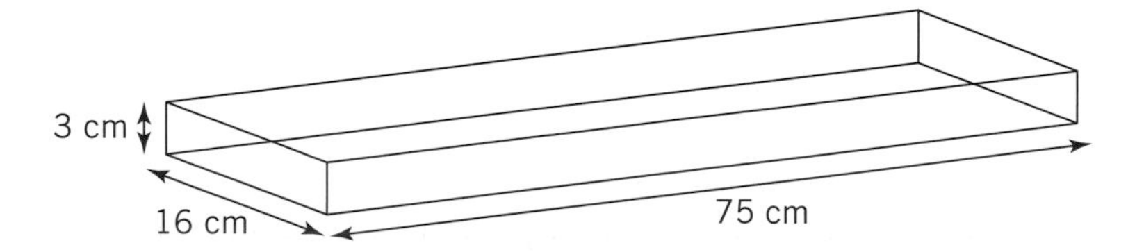

a) Calculate its volume.

b) The plank of wood has a mass of 1,800 g. Calculate the density of the wood.

c) Another block of wood has a mass of 384 g and is a cube of side length 8 cm. Is this block of wood more or less dense than the first plank of wood?

2. The picture below shows how to measure the density of an object. At first there is 35 cm^3 of water in the measuring cylinder. When the bolt is added to the measuring cylinder, the water level rises to 39 cm^3. The mass of the bolt is 29.5 g. What is the density of the bolt?

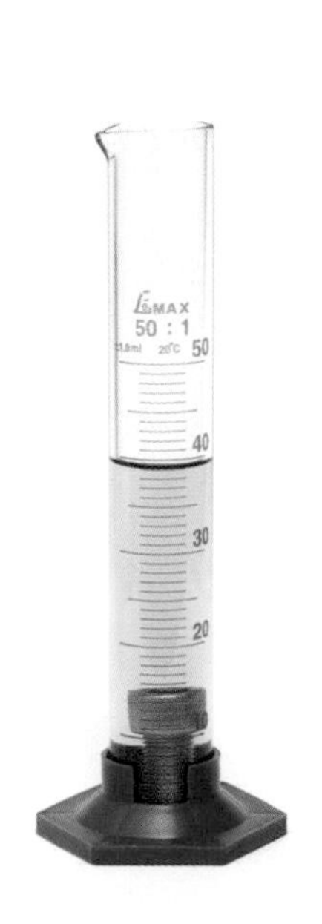

Experiment

Measuring the density of a stone

You need a measuring cylinder and a stone that will fit inside it without getting stuck.

Method

- Measure the mass of the stone.
- Half fill the measuring cylinder with water. Make sure you have enough water to completely cover the stone, but not so much that the water overflows when you put the stone in.
- Record the volume of water in the measuring cylinder.
- Gently place the stone into the measuring cylinder so that the water doesn't splash out. Record the volume of water and stone.

Questions

1. Subtract the initial volume from the final volume to find the volume of the stone.
2. Divide the mass of the stone by the volume to find the density of the stone.

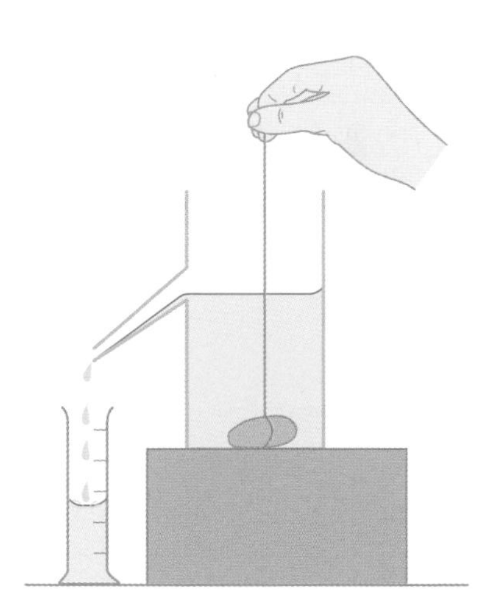

▲ A displacement can uses a similar principle: the can is filled so that water is right up to the spout. The object is added to the displacement can and displaces water, which flows down the spout and can be collected. The volume of water that is collected can be measured with a measuring cylinder and this is the volume of the solid

Why do things float?

Objects float if they are less dense than their surroundings and they sink if their density is greater. The density of water is 1 g/cm^3 and the density of air is about 0.001 g/cm^3.

1. The table gives the masses, volumes and densities of some objects.

Object	Mass (g)	Volume (cm^3)	Density (g/cm^3)
helium balloon		14,100	0.00025
log	6,800		0.8
grain of sand	0.0032	0.002	

a) By calculating the missing values in the table, find out which object is:

i) the lightest

ii) the largest.

b) Which objects will float on water?

c) Which objects will float in the air?

▲ Despite being made of metal, which is denser than water, the shape of this aircraft carrier makes its average density, which includes all the air inside it, less than that of water. The fighter planes are much denser than the air and therefore they cannot float in air. Instead, they need wings and jet engines to keep them above the ground

How can we describe matter?

Different solids can have very different properties, despite being the same state of matter. Some are very flexible such as rubber or an elastic band. These materials can be stretched and then spring back into their original shape. Other solids such as modelling clay can also be stretched but these do not return to their original shape.

When some materials are stretched, compressed or deformed and then released, they will return back to their original shape, returning all the energy that was used to deform it. Such materials are called **elastic**. Materials that can be deformed and stay that way are described as **plastic**.

ABC

An **elastic** material can return to its original shape after it is deformed.

A **plastic** material does not return to its original shape after it is deformed.

▶ Wet clay is very soft and malleable. This means it can be deformed without it breaking. When the clay is fired in a kiln, its properties change. Which terms describe fired clay?

Some materials are easy to deform and are described as soft. Others are harder to deform and are called hard or tough.

Comparing hardness

The Mohs scale is used to measure the hardness of rocks and minerals (see Chapter 10, The Earth, for more on rocks and minerals and their hardness). A mineral with a higher number on the scale can scratch a mineral with a lower number. Therefore quartz can scratch calcite, but not diamond. There are ten minerals that define the scale.

Mohs hardness	Mineral
1	talc
2	gypsum
3	calcite
4	fluorite
5	apatite
6	orthoclase
7	quartz
8	topaz
9	corundum
10	diamond

Talc is 1 on the Mohs scale

Apatite is 5 on the Mohs scale

Diamond is 10 on the Mohs scale

Find some materials, such as a stone, a nail, a coin, a piece of wood and a plastic pot. Carry out your own scratch tests to see which materials are able to scratch the others. Construct your own scale of hardness using these objects.

When a force is applied to a material, it sometimes breaks. Other materials are able to deform a lot before they break. A material that can be deformed under pressure (such as hammering it or rolling it) without breaking is described as being **malleable**. If a material can be deformed under tension and stretched into a wire, it is described as **ductile**. If a material is ductile, it is usually malleable as well. The opposite of a malleable or ductile material is one that breaks without permanently deforming. These materials are described as **brittle**.

ABC

A **malleable** material is one that can be deformed by a large amount without breaking.

A **ductile** material is one that can be stretched out into a wire.

A **brittle** material is one that will break without being deformed when a pressure is applied.

When this putty is stretched, it becomes long and thin. This behavior is described as ductile

Identifying material properties

Look at the pictures below. Which material property has been chosen when designing the object or selecting the material used? What would happen if a material with different properties were used?

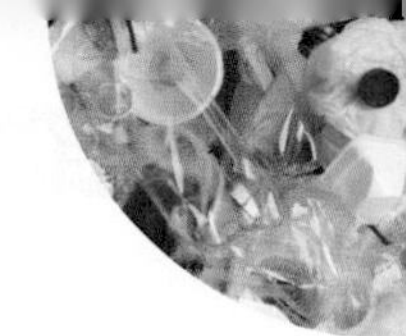

What are metals?

Elements can be arranged into the periodic table (see Chapter 7, Atoms, elements and compounds). The elements on the left-hand side of the periodic table are metals. They have different chemical properties to non-metals (see Chapter 9, Chemical reactions). They also tend to have common physical properties.

Metals often have high melting points and high boiling points, although there are some exceptions. For example, mercury has a melting point of −39°C, making it liquid at room temperature and pressure. The high melting points of most metals show that the forces between the atoms are very strong. As a result, they require a lot of energy before the forces can be overcome and the metal melts. The strong forces between the atoms also make metals difficult to deform and so most metals are hard and tough.

▲ Mercury is a metal and has many metallic properties, such as its shiny surface and ability to conduct electricity. It is unusual, however, in that it has a very low melting point and so exists as a liquid at room temperature

ATL Communication skills

Technology and collaboration

In around 3000 BC, many early people developed the technology to produce metal from metal ores. The properties of metals have many advantages over stone and this new technology marked the end of the Stone Age and the beginning of the Bronze Age.

People had to collaborate to produce metal, and the Bronze Age saw many cultures developing trade with one another. Traders needed a way to communicate without meeting face to face, and so it is probably no coincidence that many cultures developed early writing in the Bronze Age. Between 1200 and 800 BC, methods of producing iron were developed, which is a stronger metal than bronze. With this, the Iron Age began.

Since then, many new technologies have enabled people to collaborate in different ways. The electrical properties of metals allow for signals to be carried around the world. The properties of semiconductors have enabled computing to develop. This has created new ways for us to collaborate with one another, even at great distances.

▲ This is an example of cuneiform writing—one of the earliest writing systems, invented by the Sumerians during the early Bronze Age, in what is now southern Iraq

1. Give an example of an advantage of using metals instead of stone.
2. Making bronze requires copper and tin ores. Why do you think that the Bronze Age coincided with increased trade between groups of people?
3. Give an example of an advantage and a disadvantage of collaborating with others using modern technology.

Does the hardness of metals relate to their melting point?

It is suggested that the hardness of metals might be related to their melting point. The hypothesis is that metals with a low melting point are softer because they are closer to the liquid state than metals with high melting points.

The table shows the hardness of ten metals based on the Mohs scale and their melting points.

Metal	Hardness (Mohs scale)	Melting point (°C)
aluminium	2.8	660
boron	9.3	2,075
calcium	1.8	842
copper	3.0	1,085
gold	2.5	1,064
iron	4.0	1,538
lead	1.5	328
sodium	0.5	98
titanium	6.0	1,668
tungsten	7.5	3,422

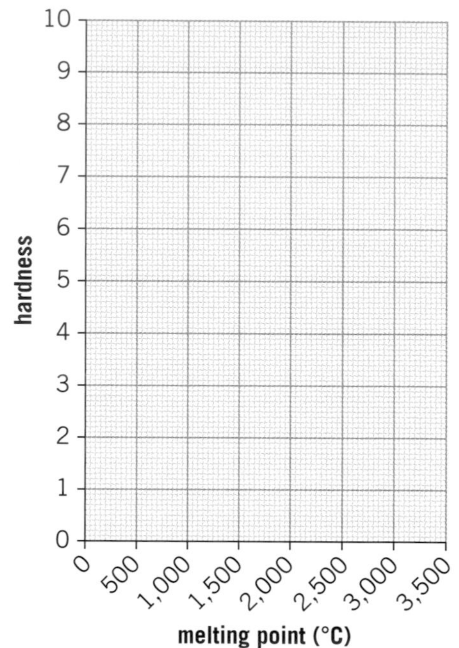

1. Plot a graph of the data with hardness on the *y*-axis and melting point on the *x*-axis. Add a line of best fit. You might want to use a copy of the axes shown above.
2. Does your graph support the hypothesis that metals with higher melting points are harder?
3. Cobalt is a metal that has a melting point of 1,495°C. Use your graph to predict the hardness of cobalt on the Mohs scale.

Metals are generally malleable and ductile, although large forces may be required to deform them. This is because the atoms in metals can form a large crystal structure, with many layers of atoms. It is possible for layers to slide past each other but still end up bound together.

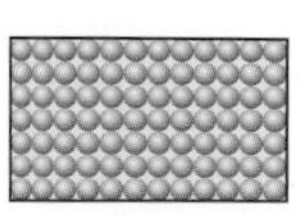

▲ Atoms in metals form a large crystal structure and the bonds between the atoms are not fixed

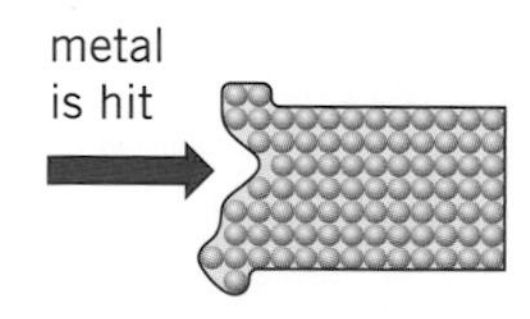

▲ When a metal is hit, it will dent, bend or deform in some way. We say that the metal is malleable

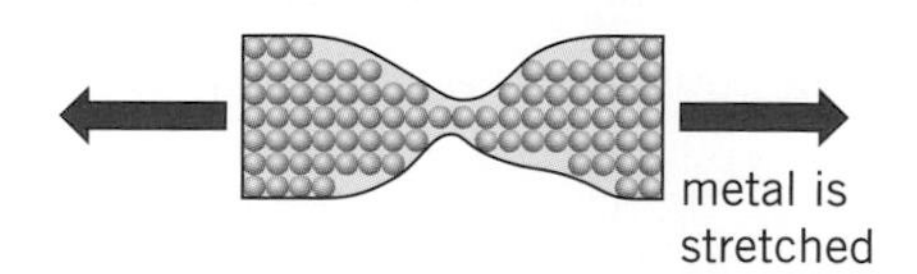

▲ If the metal is stretched, it will deform and get longer and thinner. We say that the metal is ductile

Metals also tend to be good **thermal** and **electrical conductors**. A good thermal conductor is a material that allows heat energy to be transferred through it. As a result, if a sample of a good thermal conductor is heated on one side, the thermal energy will be transferred through the material to the other side and will heat up. The opposite of a good thermal conductor is a **thermal insulator**. When one side of a thermal insulator is heated, the other side will not heat up significantly as the thermal energy is not transferred through the insulator. (See Chapter 6, Heat and light, for more on conduction.)

An electrical conductor is a material that allows a current to flow through it with little resistance. Metals tend to be good electrical conductors with copper and gold being the best. This is why the cables and wires used to carry electricity around the world and inside electrical appliances have wires with a metal core.

ABC

A **thermal conductor** is a material that allows heat energy to pass through it well.

An **electrical conductor** is a material that allows an electrical current to flow.

A **thermal insulator** is a material that does not allow heat energy to flow through it.

An **electrical insulator** is a material that does not allow an electrical current to flow.

An iron is designed to conduct heat from the heating element to the clothes that are being ironed. The base of the iron is made of metal because it is a good conductor. The handle is made from plastic so that it doesn't conduct heat energy to your hand

Experiment

Comparing thermal conductivity

For this experiment you will need a 30 cm length of copper pipe and a 30 cm length of plastic pipe that are approximately the same diameter.

Method

- Fill a large beaker with a 15 cm depth of boiling water.
- Place one end of each pipe into the water.
- Monitor the temperature of the other end of the pipes. You could tape a thermometer to the end of each pipe to help with this.

Questions

1. Which pipe conducts the heat energy fastest?
2. Which pipe is better for carrying hot water in a house?

Experiment

Investigating electrical conductivity

For this experiment you will need a 6 V battery pack, a 6 V bulb, some wires and two crocodile clips. You also need some materials to test. These could include an iron nail, a pencil lead (alternatively, use a soft pencil to shade in a rectangle on a piece of paper), a piece of copper pipe, a stick, a piece of aluminium foil, some salty water and a stone.

Method

- Assemble the circuit shown.

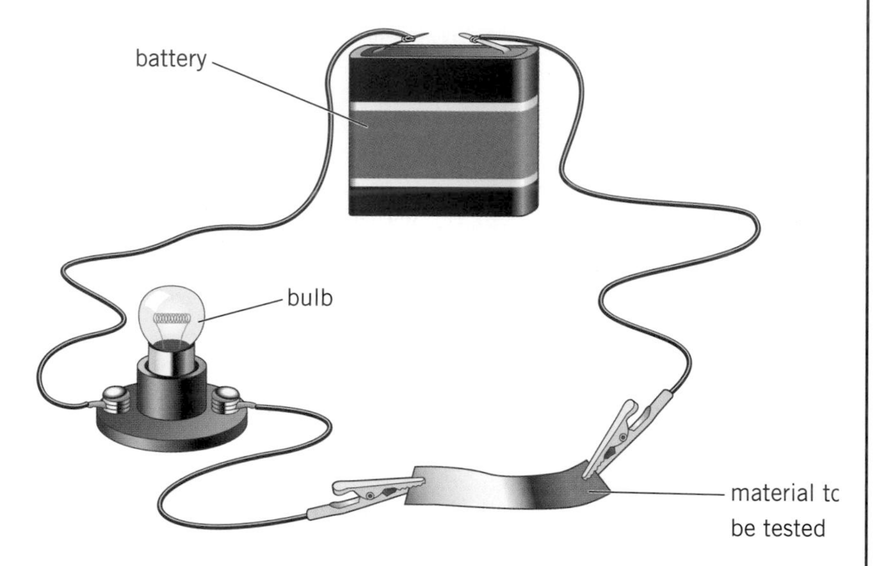

- Use crocodile clips to attach each test object into the circuit separately. (If using salty water, you can dip the ends of the crocodile clips into the water. If the object is too large to use a crocodile clip, you can hold the crocodile clip in contact with the object, one clip on either side.)

Questions

1. With which objects does the bulb light up?
2. How can you tell which materials are better conductors?

Summative assessment

Statement of inquiry:

New materials with differing properties have helped to create today's global society and may hold the answers to some of the problems of the future.

In engineering and technology, scientists want to choose the best material for a given application. The physical properties of different materials and substances will affect their suitability. In this assessment, we look at metals and their properties. We will also see some materials that might replace metals in the future.

The properties of metals

1. In which state of matter are the molecules most spread out?

A. Solid
B. Liquid
C. Gas
D. Metal

2. Which change of phase occurs when a solid metal is changed into a liquid?

A. Melting
B. Freezing
C. Vaporization
D. Sublimation

3. Metals are often described as being tough. Which of the following descriptions best matches a material described as tough?

A. It has a low boiling point.
B. It is hard to deform.
C. When it is deformed, it returns to its original shape.
D. When a force is applied, it breaks before it deforms.

4. Metals are usually good thermal conductors. Which of the following metal objects uses this property of a metal?

A. A copper wire in the power lead of an appliance
B. A metal saucepan
C. The metal head of a hammer
D. A metal cymbal

5. A hollow metal sphere has a volume of 50 cm^3. It has a mass of 40 g. It is placed in water, which has a density of 1 g/cm^3. Which statement best explains what happens?

A. The metal sphere floats because its density is less than the density of water.
B. The metal sphere sinks because its density is less than the density of water.
C. The metal sphere floats because its density is greater than the density of water.
D. The metal sphere sinks because its density is greater than the density of water.

6. Diamond is a crystal form of the element carbon. It is an excellent thermal conductor and it is the hardest naturally occurring mineral with a score of 10 on the Mohs scale. Which statement is correct?

A. Diamond is a metal.
B. Diamond could be used in an electrical circuit because it is a good conductor.
C. It is easy to scratch diamond.
D. The shape of the diamond crystal is due to the arrangement of the carbon atoms.

7. A ball has a volume of 30 cm^3. It has a density of 2 g/cm^3. What is its mass?

A. 2 g
B. 15 g
C. 30 g
D. 60 g

The hardness of metals

A pupil is researching whether there is a link between the hardness of metals and their melting points. She researches six metals that all have the same crystal structure and finds their approximate hardness on the Mohs scale and their melting points.

She finds the following data: aluminium has a hardness of 2.8 and a melting point of 660°C; copper has a hardness of 3 and a melting point of 1,085°C; gold has a hardness of 2.5 and a melting point of 1,064°C; lead has a hardness of 1.5 and a melting point of 327°C; nickel has a hardness of 4.0 and a melting point of 1,455°C; and platinum has a hardness of 3.5 and a melting point of 1,768°C.

8. Explain why it was important that all six metals had the same crystal structure. [2]
9. Put the data into a suitable table. [2]
10. Plot a graph of melting point (*y*-axis) against hardness (*x*-axis). [4]
11. Does your graph suggest that there is a relationship between these quantities? [2]

The pupil thinks that the differences in hardness and melting point might be due to how closely packed the metal atoms are. She researches the distance between the atoms in the metal lattice. She finds the following information.

Metal	Atomic spacing (billionths of mm)
aluminium	286
copper	256
gold	288
lead	350
nickel	249
platinum	277

12. Write a hypothesis for how the atomic spacing might affect the hardness of a metal. [2]
13. Produce a table of data and plot a graph to find out if your hypothesis is correct. [8]

Graphene

In 2004, Andre Geim and Konstantin Novoselov succeeded in extracting thin layers of carbon that were only one atom thick. This material is called graphene.

Graphene is a very good electrical and thermal conductor and is the strongest known material. As a result of their discovery and the unique properties of graphene, they were awarded the 2010 Nobel Prize for Physics. Later research has suggested that graphene could be used to make materials that are ten times stronger than steel, but up to 23 times less dense.

14. Describe an advantage of having a material that is ten times stronger than steel. [2]

15. Describe an application where having a material that is much less dense than steel would be useful. [2]

Today, the exciting potential of graphene in technology attracts large amounts of funding. The European Union has a research project with funding of 100 million euros ($113 million) per year. Potential uses of graphene include improvements to battery technology and more efficient solar cells.

16. Explain the implications of being able to harness some of the potential uses of graphene. [4]

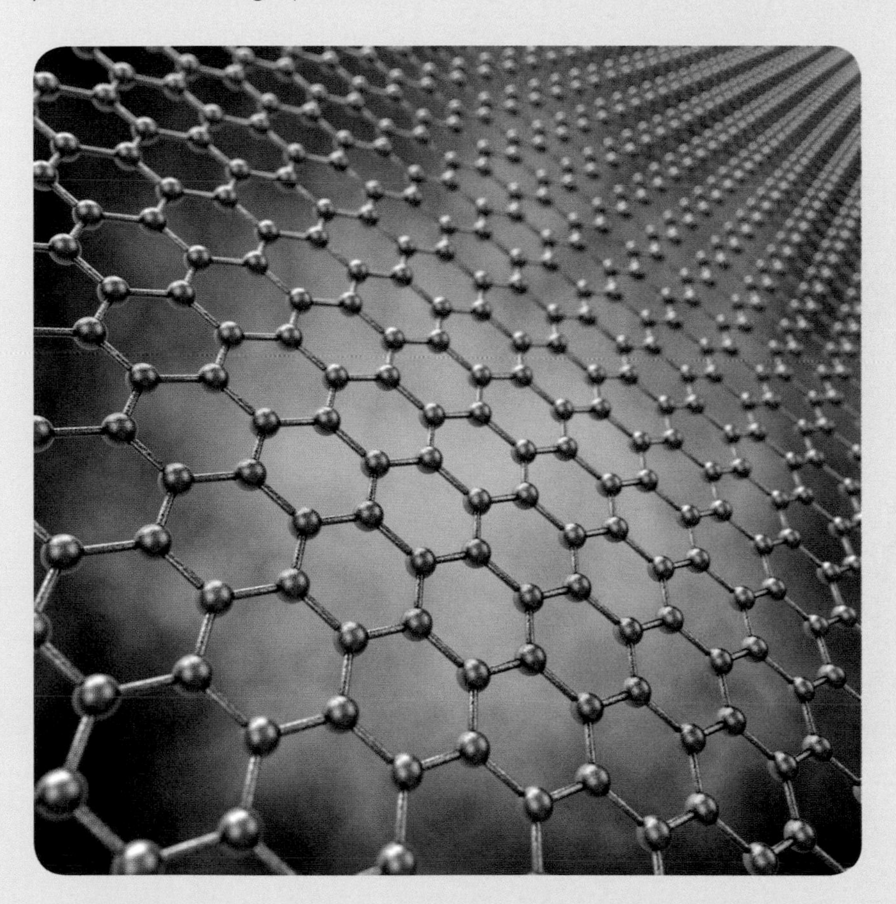

This illustration of graphene shows the carbon atoms as spheres and the bonds between them as rods. Graphene is a very thin sheet of carbon atoms, only one atom thick. It has very unusual properties such as excellent thermal and electrical conductivity, very high strength and low density

9 Chemical reactions

Key concept: Change

Related concepts: Function, Evidence

Global context: Fairness and development

◀ Chemical reactions can release energy. Often this is in the form of heat, but in the abdomen of a common glow-worm (*Lampyris noctiluca*), chemicals react to release light. The process in which living organisms can emit light is called bioluminescence. How else do living organisms use chemical reactions?

▶ These bronze tools all date from between 2000 BC to 1000 BC. Ancient civilizations learned to use chemical reactions to extract metals from rocks that contained their ores. The Bronze Age and the Iron Age were marked by the development of technology to produce and use these new materials. Which other chemical reactions are associated with milestones in human development?

▶ Chemical reactions in our muscles cause them to contract. One of the chemicals involved in this reaction is adenosine triphosphate (ATP), which releases energy and allows our muscles to do work such as climbing up a cliff. Can you think of any activities that you do that require ATP?

Statement of inquiry:

As people have used chemical reactions, their society has changed and developed.

▼ While many things in nature appear blue—the sea and the sky for example—neither of these can be used to color something else blue. The ancient Egyptians devised a method of making a pigment, called Egyptian blue, in around 3250 BC. To make it, they mixed azurite (a mineral containing copper), limestone, quartz sand and some ash, and heated it. This causes a chemical reaction to occur, which produces carbon dioxide, water vapor and a compound called calcium copper silicate, which has a blue color. Which other chemical reactions create interesting colors?

Key concept: Change

Related concepts: Function, Evidence

Global context: Fairness and development

Statement of inquiry:

As people have used chemical reactions, their society has changed and developed.

Introduction

Chemical reactions are essential to our lives. The chemical reactions involved in respiration allow living organisms to release the energy that keeps them alive. We use fuels to generate energy, enabling us to heat our homes, power machines and to provide electricity. In this chapter, we look at different chemical reactions and the factors that affect how the reactions progress. Because the changes involved in chemical reactions underpin our lives, the key concept of the chapter is change and one of the related concepts in this chapter is function.

Chemical changes are different to the physical changes such as boiling or melting that are covered in Chapter 8, Properties of matter. In a physical change, the substance may appear in very different forms. However, it is still the same substance with the same molecules containing the same types of atom. For example, if a metal such as tin is melted into a liquid, it will appear different to the solid tin metal that it was before. However, some of its properties are still the same, so the liquid tin metal will still have good thermal and electrical conductivity. To identify whether a chemical reaction has taken place, it is important to compare the properties of the initial substances (these are called the **reactants**) with the substances that result from the reaction (these are called the **products**). For this reason, the other related concept of the chapter is evidence.

Harnessing chemical reactions has transformed our lives. However, the use of chemicals can create shortages and the price of chemicals can affect their availability. Additionally, the products of chemical reactions can affect others. For this reason, the global context of the chapter is fairness and development.

ABC

The **reactants** are the chemical substances that are used in a reaction.

The **products** are the chemical substances that are produced in a reaction.

A **chemical reaction** occurs when chemical bonds are broken or formed. This means that the reactants and the products are different chemicals.

▶ Fritz Haber developed a process that converted nitrogen from the atmosphere into ammonia. Today, this is widely used to produce nitrogen fertilizer for farming. Without these fertilizers, we would struggle to produce enough food to support the global population. However, the fertilizers also cause problems. They run into rivers and disrupt these ecosystems

What is a chemical reaction?

Chapter 7, Atoms, elements and compounds, compares the difference between a mixture of elements and a compound. In a mixture, the elements are not bonded together, but in a compound, the atoms are joined together with chemical bonds. A chemical reaction occurs when chemical bonds change. As a result, the products of the reaction will contain different substances from the initial reactants.

Some chemical reactions create a product from simpler chemical compounds or even its elements. Such a reaction is often called chemical **synthesis**. Other reactions break a more complex compound into smaller product compounds—this is called **decomposition**. Often a decomposition reaction requires heat to be supplied for the reaction to occur. Such a reaction is called **thermal decomposition**.

ABC

Synthesis is a chemical reaction in which simple molecules are combined to produce a more complicated molecule.

Decomposition is a chemical reaction in which more complicated molecules are broken down to simpler molecules.

Thermal decomposition is a chemical reaction where a substance is broken into simpler molecules by heating.

Experiment

Synthesis of copper sulfate

Safety

- Wear safety glasses at all times.
- Sulfuric acid causes skin burns so gloves must be worn at all times.
- Copper oxide is very toxic to aquatic life so dispose of this chemical appropriately.

Method

- Measure 25 cm^3 of dilute (0.5 M) sulfuric acid using a measuring cylinder and pour into a beaker. Your teacher may provide this already measured for you.
- Measure between 1 and 2 g of copper oxide (1 or 2 spatulas) into a beaker. Using a spatula, add a little of this copper oxide to the acid and gently warm the acid while stirring. If the acid begins to boil, remove the heat and take care not to knock the beaker.
- When the copper oxide has dissolved, add a little more and stir until it has dissolved. Repeat until the copper oxide no longer dissolves.
- Allow the mixture to cool before filtering it to remove the remaining copper oxide.

Questions

1. What evidence is there that a chemical reaction has taken place?
2. Describe how you could obtain crystals of copper sulfate. (See Chapter 8, Properties of matter.)

▲ The green color in the water is caused by algae—tiny organisms that get their energy from photosynthesis, just as plants do. When fertilizers are overused, they are washed into rivers and ponds and can cause the algae to multiply. This can block light to other organisms in the pond, ultimately causing them to die

Experiment

Thermal decomposition of copper carbonate

Safety

- Wear safety glasses and gloves at all times.

Method

- Place a spatula of copper carbonate in a test tube.
- Heat the test tube in a Bunsen flame.

Question

1. What evidence is there for a chemical reaction taking place?

Experiment

Decomposition of water by electrolysis

Water can be decomposed into its constituent elements using an electric current. The process of using an electric current to cause a chemical reaction is called electrolysis.

For this experiment you need a 6–9 V battery pack or DC power supply. You also need two electrodes; for example, two steel nails, or two pencils sharpened at both ends.

Method

- Half fill a beaker with water.
- Place the electrodes in the water so that they are not touching.
- Connect the electrodes to the battery pack.
- Pure water is a poor conductor and so the decomposition is slow. The reaction can be made faster by adding a small amount of magnesium sulfate (a spatula is more than enough).

Questions

1. Which electrode produces more gas?
2. The chemical formula for water is H_2O, which means that each molecule of water has two hydrogen atoms and one oxygen atom. Use this to identify which electrode is producing each gas.

What evidence suggests a chemical reaction has taken place?

A chemical reaction takes place on a tiny scale with changes in the bonding between atoms. It is not possible to observe this directly. However, there are often clues that might suggest that a chemical reaction is taking place.

- **Color changes:** a change in color can mean that a new substance has formed.
- **Gas released:** if a gas is released this might show that a chemical reaction has taken place. In liquids look for bubbles appearing, but remember that it might be because the liquid is boiling.
- **Smells:** a smell that was not there before might show that a new chemical substance has formed. Do not inhale chemicals, however, as they may be harmful or poisonous.
- **Precipitate forms:** when two liquids are mixed, a solid sometimes forms. This is called a **precipitate**. It shows that a chemical reaction has taken place.
- **Temperature changes:** many chemical reactions require energy or release it. This causes the temperature to decrease or increase.

1. What evidence is there that a chemical reaction has taken place in each of these situations?
   a) An egg frying
   b) Wood burning
   c) An apple rotting

ABC When a chemical reaction between two solutions produces an insoluble product, the product forms as a solid that is suspended in the solution. This is called a **precipitate**.

Experiment

Evidence for chemical reactions

Safety

- Wear safety glasses and gloves at all times.
- All experiments must be conducted within a fume cupboard.
- Lead nitrate is very toxic to aquatic life so must be disposed of appropriately.

Method

Carry out the following reactions using the dilute solutions provided by your teacher. Mix about 10 cm^3 of each reactant in a conical flask. Observe the reactions and record any evidence for a reaction taking place.

- Lead nitrate + potassium iodide
- Ammonium chloride + sodium hydroxide
- Sodium hydroxide + hydrochloric acid
- Sodium carbonate + hydrochloric acid

What causes chemical reactions?

Chemical reactions occur when bonds between atoms are made or broken. It takes energy to break bonds between atoms, and energy is released when bonds are formed. This is why most decomposition reactions require heat to be supplied for a compound to be broken into two or more simpler molecules. More energy is required to break the initial chemical bonds than is released in forming the new chemical bonds. This type of reaction, which requires more energy to be supplied than it releases, is called **endothermic**.

ABC An **endothermic** reaction is a reaction that requires heat energy to be supplied or takes energy from its surroundings.

An **exothermic** reaction is a reaction that results in an overall release of energy to its surroundings.

The opposite of an endothermic reaction is an **exothermic** reaction. An exothermic reaction may require some energy to be supplied to break the existing chemical bonds in the reactants. However, more energy is released by forming the chemical bonds in the product.

Experiment

Identifying endothermic and exothermic reactions

Safety

- Wear safety glasses and gloves at all times.

Carry out these experiments. Use a thermometer to see if you can detect any temperature change.

a) Put acetic acid (vinegar) in a plastic cup to a depth of 1 cm. Add a spatula of sodium hydrogen carbonate (bicarbonate of soda) to it.

b) Put water in a plastic cup to a depth of 1 cm. Add two spatulas of calcium chloride to it.

Questions

1. Is each reaction endothermic or exothermic?

2. What evidence is there to suggest that a chemical reaction has taken place?

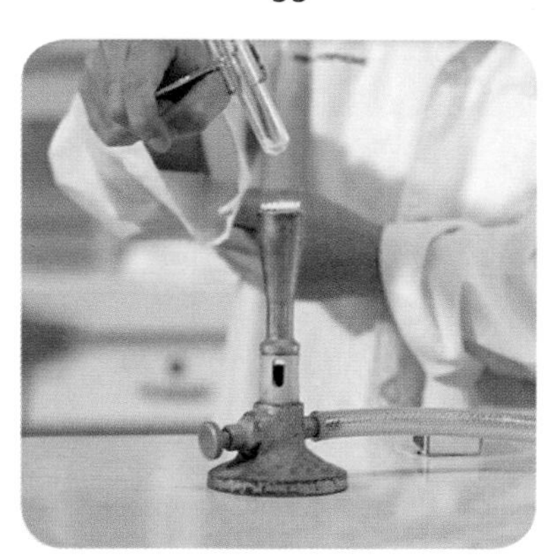

▲ Many chemical reactions require heat to be supplied for them to happen. These reactions are endothermic. A common way of supplying this energy is using a Bunsen burner. The chemical reactions that occur when the gas burns in the Bunsen burner are exothermic—they release energy

What are explosions?

Some reactions release a large amount of energy—they are highly exothermic. If they do this very quickly, then they may cause an explosion.

▶ Explosions are highly exothermic reactions. The heat produced causes gases to expand rapidly and this blasts material outwards with a large force

Nitrogen triiodide

Nitrogen triiodide is a very unstable compound and decomposes into its constituent elements, nitrogen and iodine. Use the internet to find a video of its decomposition.

1. What is required to make nitrogen triiodide decompose?

2. Is this reaction endothermic or exothermic?

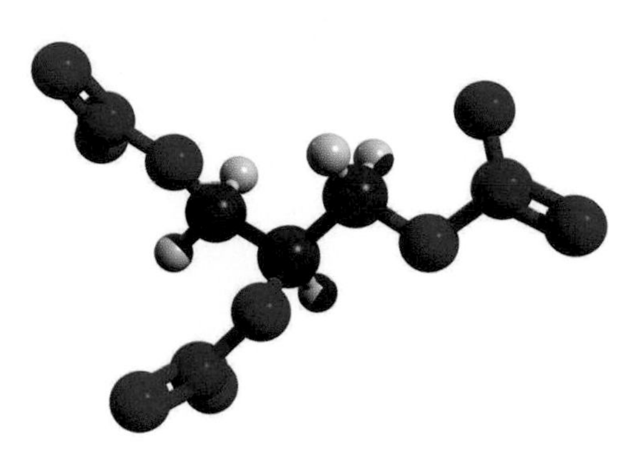

▲ Nitroglycerin is an explosive that was discovered in the mid-nineteenth century by chemist Ascanio Sobrero. As an explosive, it decomposes with a highly exothermic reaction. As with other exothermic reactions, some energy is required to break the initial chemical bonds and start the reaction. In the case of nitroglycerin, this energy could be supplied as a physical force such as a shock or even a loud sound. This makes nitroglycerin highly unstable and very dangerous

▲ One of Sobrero's pupils was Alfred Nobel, who experimented with making nitroglycerin safer to use, after his brother was killed in a nitroglycerin explosion. His company patented the explosive dynamite, which was stable enough to use but was still highly explosive. Later, dynamite was used in the Crimean war and Nobel was concerned that he would be remembered for causing deaths rather than for the peaceful uses that dynamite had, such as in mining. As a result, Nobel made a will that left all his estate to founding the Nobel prizes. These prizes are given for advances in chemistry, literature, peace, physics and physiology or medicine

Exothermic reactions

1 g of each of the elements with atomic numbers 11 to 17 was burned in chlorine gas (see Chapter 7, Atoms, elements and compounds, for more on atomic number and elements). In each case there was an exothermic reaction and the amount of energy released was measured. The results are shown in the table.

▲ Sodium metal burns in chlorine gas with a yellow flame

Atomic number	Energy released (kJ)
11	18
12	27
13	26
14	23
15	
16	3
17	0

1. Plot a suitable graph of these results.
2. No measurement was made for the element with an atomic number of 15. Use your graph to predict the amount of energy released.
3. Why was it important that 1 g of each element was burned in chlorine?
4. The element with atomic number 17 did not react with chlorine gas. Suggest why this is.

ATL **Communication skills**

Choosing a line of best fit

Graphs are a good way of presenting data in a visual form and a line of best fit can help to show the trend.

Once you have drawn a graph, consider whether there is an obvious trend.

If the trend appears to be linear (the points all seem to lie near a straight line), then you should draw a straight line of best fit. Use a ruler and draw the line that passes as close to the points as possible. The line should have approximately the same number of points on either side of it. The line does not have to pass through all the points—in fact, it might not pass through any of them.

Sometimes it is clear that the data points form a curve. We call this trend non-linear. In this case, a free hand curve can be drawn.

Sometimes, there is not enough information to reliably draw a line of best fit. It might be that the trend is not clear or that there is not intended to be any trend. Sometimes you will see this type of graph with the points joined with straight lines.

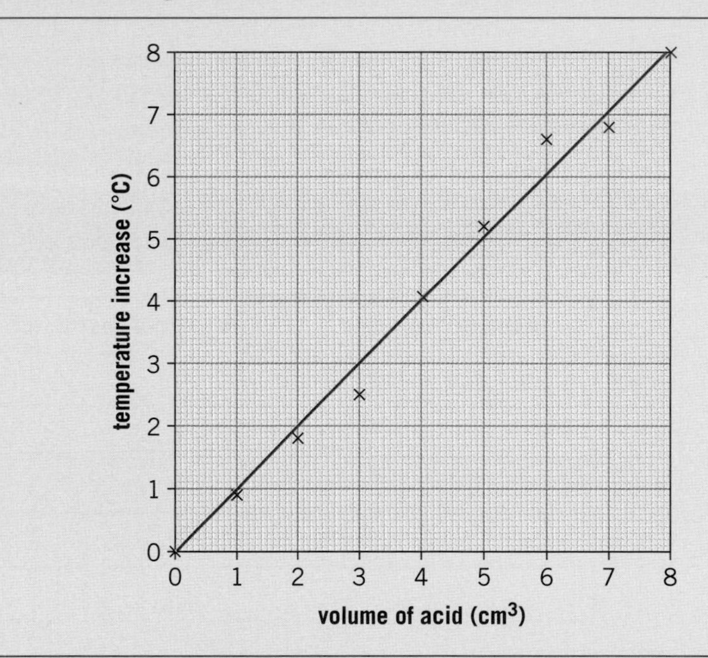

This graph shows the increase in temperature when different amounts of hydrochloric acid were added to sodium hydroxide solution.

The trend appears to be a straight line and so a straight line of best fit is drawn through the points.

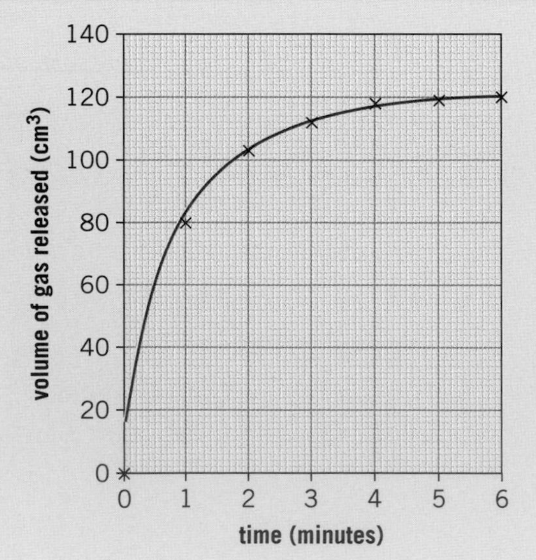

This graph shows the amount of carbon dioxide gas released from a chemical reaction with time.

The amount of carbon dioxide increases at first, but as the chemical reaction progresses, it slows down until it is complete. This gives the graph a curved trend (non-linear). Therefore a curved line of best fit is suitable for this graph.

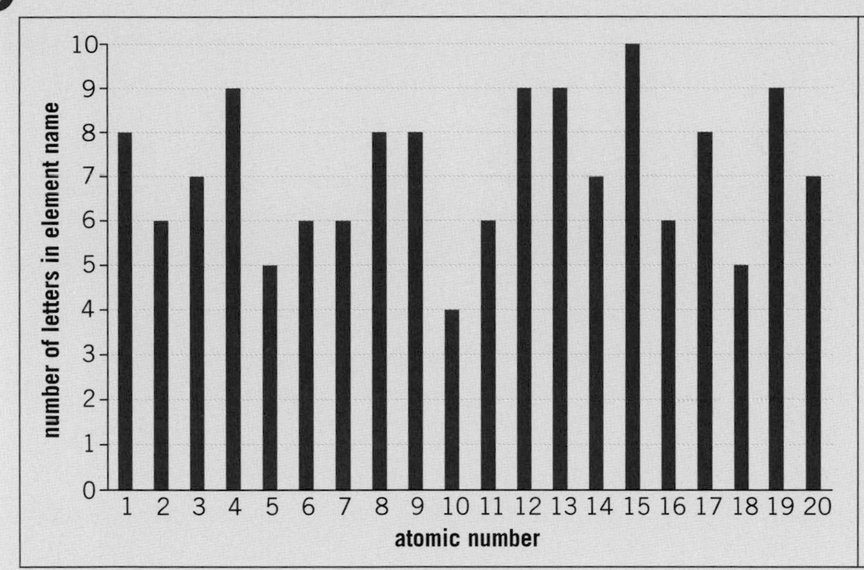

This graph shows the number of letters in the name of each of the first 20 elements plotted against the atomic number of the element.

Because atomic number is a discrete variable (it is impossible to have a half value), a bar chart has been plotted.

There is no clear trend and so no line of best fit is drawn.

How can we describe chemical reactions?

Chemical reactions can be described using word equations. In a word equation, the reactants go on the left-hand side and the products go on the right-hand side. So, the reaction of magnesium burning in oxygen to form magnesium oxide is written as:

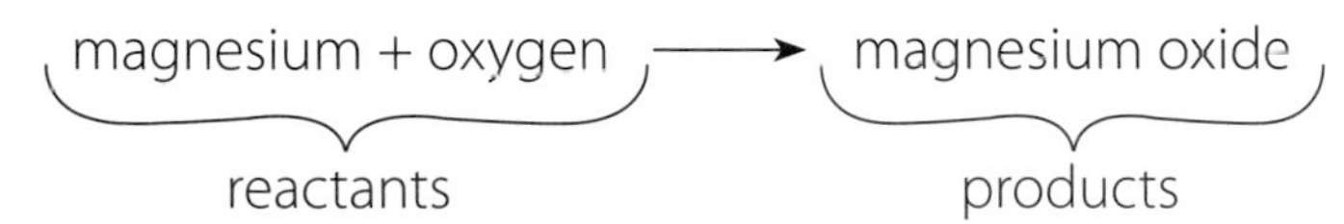

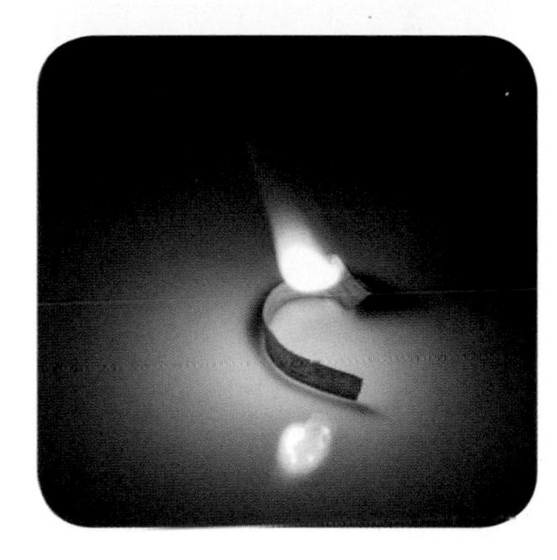

▲ Magnesium burns in oxygen to produce magneisum oxide

1. Identify the products and reactants in the following reactions. Write the reactions as word equations.
   - a) Sodium reacts with water to form hydrogen and sodium hydroxide.
   - b) Water is formed by reacting hydrogen and oxygen.
   - c) Copper oxide neutralizes hydrochloric acid to form copper chloride and water.

What are acids?

ABC An **acid** is a substance that has a pH less than 7.

There is an important class of chemicals called **acids**. You may be familiar with citric acid, which occurs in foods (particularly citrus fruits such as lemons), or acetic acid, which is present in vinegar. Acids that are more common in the laboratory are sulfuric acid, hydrochloric acid and nitric acid.

Many acids are corrosive, which means that they can erode surfaces and wear them away through chemical reactions. This can make acids dangerous if they get on your skin or in your eyes. However, not all acids are dangerous. Lemon juice and vinegar will hurt if you splash them on a cut or a small amount gets in your eyes, but they will not cause damage. This is because they are not very acidic and the acid is diluted with enough water to make them safe.

The dilute acids that you might use in the laboratory (for example, dilute hydrochloric acid or sulfuric acid) are a bit more acidic and you should wear eye protection whenever you use them in experiments. Concentrated acids would only be used by experienced chemists as these can be very **corrosive**.

Some substances can neutralize acids and are thought of as the opposite of acids. These substances are called **bases**. Some bases dissolve in water and are called **alkalis**. The most commonly used alkaline solutions in the laboratory are sodium hydroxide and ammonium hydroxide. Strongly alkaline solutions can also be corrosive.

Acids and alkalis can be measured on the **pH** scale, which runs from 0 to 14. The mid-point of the scale, a pH of 7, represents a solution that is neither acidic nor alkaline, such as pure water. A pH lower than 7 is acidic. The lower the pH is, the more acidic the solution is. If the pH is above 7, the solution is alkaline and the higher the pH, the more alkaline the solution is.

pH can be tested with an **indicator**. There are many different indicators, but two commonly used indicators are universal indicator solution and universal indicator paper. To use them, either add a couple of drops of universal indicator, or dip a small piece of universal indicator paper into the solution. The color that the paper or the solution turns can be compared with a scale to show the pH.

ABC

A **base** is a substance that has a pH greater than 7. Bases neutralize acids.

An **alkali** is a base that is soluble in water.

A **corrosive** substance is a substance that eats away and erodes other materials.

An **indicator** is a chemical compound that changes color depending on the pH of its surroundings.

▼ Adding a few drops of universal indicator to a solution, or dipping a piece of universal indicator paper will show how acidic a solution is

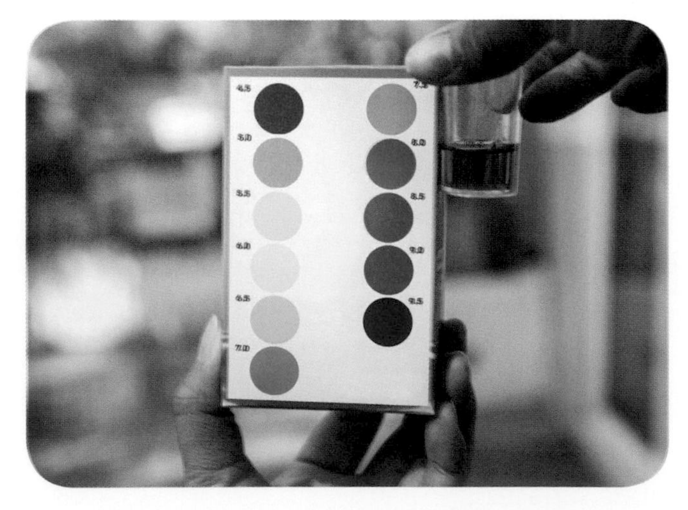

When an acid reacts with a base, the reaction is called a **neutralization reaction**. This type of reaction will produce water and another substance that is called a **salt**.

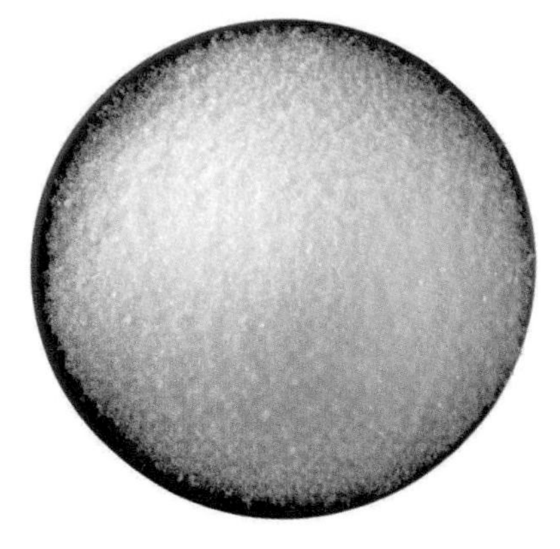

▲ Table salt (sodium chloride, on the left) used in cooking is only one example of a type of substance that chemists call a salt. These pictures also show copper sulfate (middle) and magnesium sulfate (right). In chemistry, all these compounds would be called a salt. While copper sulfate has a blue color, magnesium sulfate appears very similar to sodium chloride. It is important that you do not assume a substance that looks like salt is safe to eat

The salt that is used in cooking, sodium chloride, is a good example of a substance that chemists call a salt—but it is only one example. A salt is formed when an acid is neutralized. As a result, it is a compound that has two parts—one from the acid and one from the base. For example, if hydrochloric acid is used to neutralize sodium hydroxide, the products are sodium chloride and water. However, if sulfuric acid is used instead of hydrochloric acid, the result would have been sodium sulfate. The type of acid will determine the type of salt that is produced. Sulfuric acid produces sulfates, hydrochloric acid produces chlorides and nitric acid will produce nitrates.

The other part of the salt's name comes from the base or alkali. In the example above, sodium hydroxide was used and the salt that was produced was sodium chloride or sodium sulfate. If a different base had been used, copper oxide for example, then copper chloride or copper sulfate would have been produced.

These reactions can be summarized in the following equations:

sodium hydroxide + hydrochloric acid → sodium chloride + water

sodium hydroxide + sulfuric acid → sodium sulfate + water

sodium hydroxide + nitric acid → sodium nitrate + water

copper oxide + hydrochloric acid → copper chloride + water

copper oxide + sulfuric acid → copper sulfate + water

1. Identify which salt is produced from the following reactants. The first one has been done for you.

Acid	**Base**	**Salt**
hydrochloric acid	sodium hydroxide	sodium chloride
hydrochloric acid	calcium hydroxide	
sulfuric acid	ammonium hydroxide	
nitric acid	copper oxide	

ABC

A **neutralization reaction** is a reaction where an acid reacts with an acid or base to form a salt and water.

pH is a measure of how acidic or alkaline a solution is. A pH of 7 is neutral, less than 7 is acidic and greater than 7 is alkaline.

A **salt** is a chemical compound formed from the reaction of an acid with a base.

▲ Limescale is a build-up of calcium carbonate. It can block pipes

2. Limescale is a hard deposit that can block pipes. Limescale is mainly formed from calcium carbonate, which is a base. Limescale remover contains an acid, often hydrochloric acid, which reacts with the limescale to form calcium chloride. Water and carbon dioxide are also produced.
   a) Write a word equation for this reaction.
   b) Which of the products is a salt?
   c) Which of the products is a gas?
   d) What would have been produced if sulfuric acid had been used instead of hydrochloric?

How do metals react?

Chapter 8, Properties of matter, looks at how elements can be divided into metals and non-metals. As well as having different physical properties, metals and non-metals also have different chemical properties.

Metals can react with the oxygen in the air. They react to form a metal oxide. An example of this is iron, which reacts with oxygen in the air to form iron oxide; this is known as rust. The equation for this reaction is:

$$\text{iron} + \text{oxygen} \rightarrow \text{iron oxide}$$

This kind of reaction is called **oxidation**.

An **oxidation** reaction is one in which one substance gains oxygen to form a compound.

When a metal is **tarnished**, it loses its shininess (or luster).

▲ Iron can be oxidized to form rust, a reddish-brown substance shown on this rusty padlock and chain

Experiment

Rusting in different environments

You are going to create three different environments to investigate the requirements for rusting. You will need three nails that are made of iron or steel. They shouldn't be coated in any way. Ensure that the nails are clean; they are sometimes sold with a small amount of oil on them. It is a good idea to rub them with emery paper to ensure that there is nothing on the surface.

You can use conical flasks with bungs for this experiment, or you could use jar with a lid.

Method

The first environment will contain only air. Ensure the conical flask is dry. You could do this by warming it or using a small pack of silica gel (often found the packaging of new electronic items). Place an iron nail in the flask and seal it with a bung.

The second environment will have only water. Boil some water to remove the gases that are dissolved in it and let it cool a little. Put the iron nail into the flask and add enough water to cover it. Gently pour a thin layer of oil over the surface of the water so it is covered. This will stop the air from dissolving in the water again. Seal this flask with a bung.

The third environment will have water and air. Place an iron nail in the flask and add water so that the nail is half covered. Seal the flask with a bung.

Leave all three for a few days to see which environments cause the nail to rust.

Questions

1. Which environments caused the iron to rust?
2. What conditions cause iron to rust quickest?

Experiment

The effect of salt on the rate of rusting

When sodium chloride dissolves in water, it can speed up the process of rusting.

You will need eight boiling tubes or jars, and eight small iron or steel nails.

Method

- Ensure the iron nails are clean—nails are sometimes sold with a thin coating of oil to stop them rusting. A rub with emery paper or fine sandpaper can help.
- Place 10 cm^3 of distilled water in each boiling tube or jar. If distilled water is unavailable, use boiled water.
- Add salt to each boiling tube so that they have 0, 0.5, 1, 1.5, 2, and 2.5 g of salt. Label each boiling tube.
- Add an iron nail to each boiling tube and leave for a few days until some of them show signs of rusting.

Questions

1. Was there a difference in the amount of rusting?
2. Which boiling tube had the most rusting?
3. Sea water has about 0.35 g of salt per 10 cm^3 of water. Do your results suggest that iron would rust faster in sea water than in water with no salt?

▲ Does the salt in seawater increase the rate at which this wreck is rusting?

▲ Fireworks and sparklers often use magnesium because it burns with a sparkling white light

Iron reacts with oxygen in the air very slowly, but some metals react much more readily. Magnesium, for example, burns very vigorously in air with a bright flame. It is used in fireworks and flares.

Other metals also react vigorously with oxygen. Calcium, lithium, sodium and potassium all burn vigorously. Metals such as zinc and aluminium react less quickly with oxygen. Metals such as copper, lead, and mercury react very slowly and **tarnish**, but this reaction can be sped up by heating. Some metals such as silver and gold are very unreactive and do not react with the oxygen in the air.

1. Why do you think silver and gold are used for decorative purposes such as jewelry?

How do metals react with water?

If some common metals were ranked in order of their reactivity with oxygen, then the order would be:

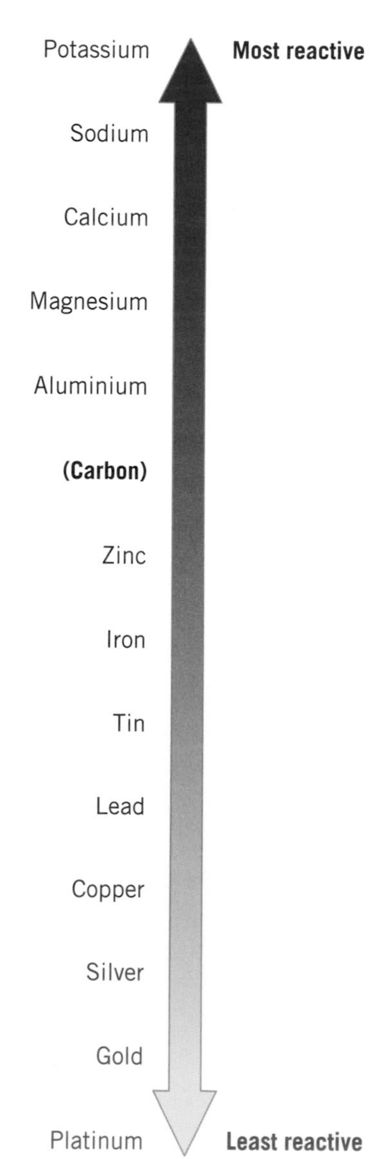

▲ This picture show potassium metal burning in oxygen

The reactivity of these metals with air is similar to their reactivity with water. Lead, copper, silver and gold are sufficiently unreactive that they do not react with water. This is why lead and copper were often used for water pipes. Indeed, the word for a plumber comes from the Latin word for lead (*plumbum*). Lead was found to be harmful to health and copper is expensive, so plastic is now more common.

Iron, zinc, aluminium and magnesium do not readily react with water unless the water is very hot (usually as steam). Metals higher up the list react with cold water. When calcium is added to water, it produces bubbles of hydrogen gas in the following reaction.

$$\text{calcium} + \text{water} \rightarrow \text{calcium hydroxide} + \text{hydrogen}$$

Similar reactions happen with sodium and potassium, except that these reactions are much more vigorous. Because of the metals' low density, sodium and potassium also float on water. Potassium reacts so vigorously that the hydrogen gas produced even ignites.

▲ Sodium and potassium react vigorously with water, releasing hydrogen gas that can ignite

The reactions of metals with water

The table gives the amount of energy released when 1 g of a metal reacts with water and the approximate time it takes for the reaction to happen.

Metal	Energy released when 1 g reacts with water (J)	Approximate time taken for reaction (seconds)
potassium	6,800	7
sodium	8,000	30
calcium	10,750	150
magnesium	14,600	430,000

1. Plot a bar chart to show the energy released by each metal.
2. Why should the data in the table be plotted as a bar chart?
3. The approximate time for the reaction of 1 g of magnesium with water is given in seconds. Express this time in a more convenient unit.
4. Although potassium is higher up the reactivity series than magnesium, it does not release as much energy in its reaction with water. Explain why the reaction of potassium with water appears more vigorous.

▲ Magnesium reacts with water to release bubbles of hydrogen gas

Rubidium and caesium

Your teacher may be able to show you the reactions of sodium and potassium with water. If not, use the internet to find videos of how they react with water.

Rubidium and caesium are metals that are even more reactive than sodium and potassium. As a result, they are very dangerous and also very expensive. Use the internet to find a video of rubidium or caesium reacting with water. How do the reactions differ from those of sodium and potassium?

How reactive are metals?

In the previous section, we saw that different metals react with oxygen and water in different ways. Metals such as potassium and sodium are the most reactive. Some metals, like iron, react slowly and others, such as silver and gold, do not react at all. This order of reactivity is called the reactivity series.

ABC A **displacement reaction** is a reaction where a more reactive metal replaces a less reactive metal in a compound.

The reactivity series also determines how metals (and compounds that contain these metals) will react with each other. If a metal is more reactive, there is more energy available when it reacts to form a compound. Because of this, a more reactive metal can be used to displace a less reactive metal from a compound. This kind of reaction is called a **displacement reaction**.

For example, when copper metal is placed in a solution of silver nitrate, the copper displaces the silver. This is because copper is more reactive than silver. The equation for this reaction is:

copper + silver nitrate → copper nitrate + silver

When copper metal is added to a solution of silver nitrate, the copper displaces the silver because it is more reactive. Silver metal is deposited as crystals on the copper metal and the solution turns a blue color because of the copper nitrate that is formed

	Metal	Reaction with oxygen	Reaction with water
most reactive	potassium	oxidize rapidly in air and burn vigorously	react with cold water
	sodium		
	calcium		
	magnesium		react with steam
	aluminium	react with oxygen in the air	
	zinc		
	iron		
	lead	react with oxygen when heated	do not react
	copper		
	silver	do not react	
least reactive	gold		

▲ The reactivity series of metals

Experiment

Comparing reactivity

Safety

- Wear safety glasses and gloves at all times.
- Copper sulfate is very toxic to aquatic life so must be disposed of appropriately.

You will need samples of copper foil, magnesium ribbon, and iron or steel (possibly a nail) and solutions of copper sulfate, magnesium sulfate and iron sulfate.

Method

- First, test the copper sulfate with the iron nail. Pour copper sulfate solution into a boiling tube to a depth of 2 cm and add the iron nail. Observe any reaction that takes place.
- Now test each solution with each metal in the same way. You will need to use a fresh solution each time.
- Record your observations and record whether a reaction takes place in a table like this.

	Copper	Iron	Magnesium
Copper sulfate			
Iron sulfate			
Magnesium sulfate			

Questions

1. Which metal was the most reactive?
2. Which metal was the least reactive?
3. Write the chemical equation for any reactions that you observe.

Displacement reactions can be used to sort the metals into a series of reactivity. The most reactive metals are potassium and sodium. The least reactive metals are silver and gold.

Metal ores

Sometimes hydrogen and carbon are placed on the reactivity series, even though they are not metals. This is because they can take part in displacement reactions. Hydrogen appears above copper and carbon appears above iron. As a result, carbon can be used to displace iron from its ore in the reaction:

carbon + iron oxide → iron + carbon dioxide

This reaction needs a high temperature, about 1,000°C. Copper, on the other hand, can be produced at a much lower temperature (about 400°C).

1. Bronze is a metal that contains copper. Ancient history is often divided into the Stone Age, the Bronze Age and the Iron Age. Although these periods started at different times in different places, the Bronze Age started around 3000 BC and the Iron Age in around 1200 BC. Why do you think that the Bronze Age came before the Iron Age?
2. What advantages did people who had the technology to produce iron have over those who did not?
3. Research the differences in the properties of iron and copper. Why is iron a more widely used material?
4. Today, iron and copper have different prices. Much of this is due to the abundance of iron and copper ores. Research the current cost of scrap iron and scrap copper. You may even be able to find out how the prices have changed in recent years. What does this tell you about the abundance of iron and copper ores?

▶ This is a reconstruction of a primitive iron furnace called a bloomery. Iron ore is heated in a charcoal fire. As well as producing the high temperatures that are required for the reaction, the charcoal is a source of carbon, which is higher up the reactivity series than iron. As a result, the carbon is able to displace the iron from its ore, producing iron metal

Summative assessment

Statement of inquiry:

As people have used chemical reactions, their society has changed and developed.

In this assessment, we will look at chemical reactions and how they can be used to make fertilizers. We will see how the use of fertilizers has changed over time and how farmers benefit from the availability of these fertilizers.

Chemical reactions

1. Which metal appears highest on the reactivity series?

A. Gold
B. Iron
C. Potassium
D. Zinc

2. A solid powder is added to water. Which of the following might be evidence that a chemical reaction has taken place?

A. The solid dissolves.
B. The solid does not dissolve.
C. Bubbles form.
D. The water evaporates.

3. A set of experiments are carried out to see which metals displace other metals from salts. The results are shown in the table. Using these results, deduce the order of these metals in the reactivity series, starting from the least reactive.

Metal	Metal salt			
	W	**X**	**Y**	**Z**
W	no reaction	reaction	no reaction	reaction
X	no reaction	no reaction	no reaction	no reaction
Y	reaction	reaction	no reaction	reaction
Z	no reaction	reaction	no reaction	no reaction

A. XZWY
B. YWXZ
C. YWZX
D. YZXW

When sodium hydroxide reacts with hydrochloric acid, sodium chloride and water are produced.

4. Which term best describes the water in this reaction?

A. Reactant
B. Product
C. Salt
D. Acid

5. Which word describes the type of reaction in the previous question?

A. Decomposition
B. Endothermic
C. Neutralization
D. Synthesis

6. Copper sulfate can be produced by reacting copper carbonate with sulfuric acid. Water and carbon dioxide are also produced. Which of these compounds is a salt?

A. Carbon dioxide
B. Copper oxide
C. Copper sulfate
D. Sulfuric acid

7. In a displacement reaction, zinc metal is added to a solution of copper nitrate. Which of the following is one of the products?

A. Copper sulfate
B. Nitric acid
C. Nitrogen gas
D. Zinc nitrate

8. When 1 g of iron filings are added to a colorless solution of lead nitrate, a yellowy brown solution forms. What is this solution?

A. Iron nitrate
B. Lead
C. Nitrogen
D. Rust

Making and testing fertilizers

Fertilizers are chemical substances that provide important substances to help plants to grow. The three most common nutrients that are supplied to plants are potassium, phosphorus and nitrogen. Other elements such as magnesium, calcium and sulfur are also used in smaller quantities.

A student proposes to make some potassium nitrate. She uses the following method. First, she reacts a small amount of potassium metal with water. This creates an alkaline solution of potassium hydroxide. She then adds a small amount of dilute nitric acid to create potassium nitrate.

9. Give two safety precautions that she should take when carrying out this method. [2]

To investigate other fertilizers, the student dissolves some magnesium sulfate in distilled water. She adds a small amount of calcium metal to the solution of magnesium sulfate and produces calcium sulfate, which is insoluble.

10. Describe how she might separate the calcium sulfate from the mixture. [2]

11. Explain why adding magnesium metal to calcium sulfate would not displace the calcium. [1]

The student decides to test the fertilizers that she has made. She decides to put some of each fertilizer (potassium nitrate, magnesium sulfate and calcium sulfate) on a separate area of grass and measures how much the grass grows in a week.

12. Suggest two control variables for this experiment [3]

13. Suggest how she might present her results. [2]

The cost and use of fertilizers

The table gives the yearly price of ammonium nitrate and the amount of ammonium nitrate that was used to fertilize crops in the USA from 2000 until 2014.

Year	Price ($ per kg)	Amount used (millions of kg)
2000	0.21	1,541
2002	0.21	1,424
2004	0.29	1,386
2006	0.40	874
2008	0.56	789
2010	0.44	653
2012	0.63	766
2014	0.62	774

Source of data: United States Department of Agriculture

14. Plot a graph with the yearly price of ammonium nitrate on the *y*-axis, and year on the *x*-axis. Plot a second graph, with the amount of ammonium nitrate used on the *y*-axis, and the year on the *x*-axis. [4]

15. Explain whether a line of best fit is appropriate for these graphs. [1]

16. Describe the trends of these two graphs. Is there a connection between the two trends? [3]

17. In which year was the most money spent on ammonium nitrate as a fertilizer in total? [2]

The availability of fertilizers

Read the following text and answer the questions that follow:

> Ammonium nitrate (NH_4NO_3) is a compound that is high in nitrogen and is often used as a fertilizer for crops. It is synthesized by reacting ammonia with nitric acid. This reaction is highly exothermic.
>
> The cost of ammonium nitrate is approximately \$300 per tonne and about 0.3 tonnes of fertilizer is used per hectare. This can increase the amount of maize that can be grown by about 3 tonnes per hectare. Maize can sell for about \$180 per tonne.
>
> However, ammonium nitrate is highly soluble. As a result, it can wash off the land and pollute rivers where it causes algae to grow quickly and block the light for other organisms.
>
> Many developing countries such as those in sub-Saharan Africa have crop yields that are up to half that of other countries. This is in part due to the limited availability of fertilizers.

18. Write a word equation for the synthesis of ammonium nitrate. [2]

19. Explain what is meant by a "highly exothermic" reaction and explain why such a reaction needs to be carried out in a safe and controlled way. [2]

20. A farmer in a developing country has 2 hectares of land, which is used for growing maize. Calculate:

a) the cost of the fertilizer that is needed [1]

b) the increase in the number of tonnes of maize that can be produced [1]

c) the total price that the extra maize can be sold for [1]

d) the profit that the farmer can make from using the fertilizer. [1]

21. Other than the increased profit to the farmer, give one advantage of the increased yields in a developing country. [1]

22. Give one disadvantage of using the fertilizer. [1]

10 The Earth

Key concept: Change

Related concepts: Transformation, Development

Global context: Fairness and development

Volcanoes are a natural hazard. The Roman city of Pompeii was built near to a dormant volcano— Mount Vesuvius. The inhabitants were used to small earthquakes happening, but in the year AD 79, Mount Vesuvius erupted and buried the city in hot ash and pumice. This preserved the city for many centuries. Which other cities are built near natural hazards?

Humans dig many resources from the Earth. Rock and sand is used for building. Metal ores are dug to be converted into metals and coal is dug for fuel. In doing so we change the surface of the Earth. Which other processes change the natural landscape?

Statement of inquiry:

The Earth has been transformed by slow processes that still continue today.

▲ When rock is subjected to the extreme temperatures and pressures that are found deep in the Earth, it changes form. As a result, the same constituents that are found on the Earth's surface are transformed into new types of rock. These rocks are uncut diamonds and they are made from carbon, which is a common element found in rocks. When they are cut, they will command a high price. What makes them expensive?

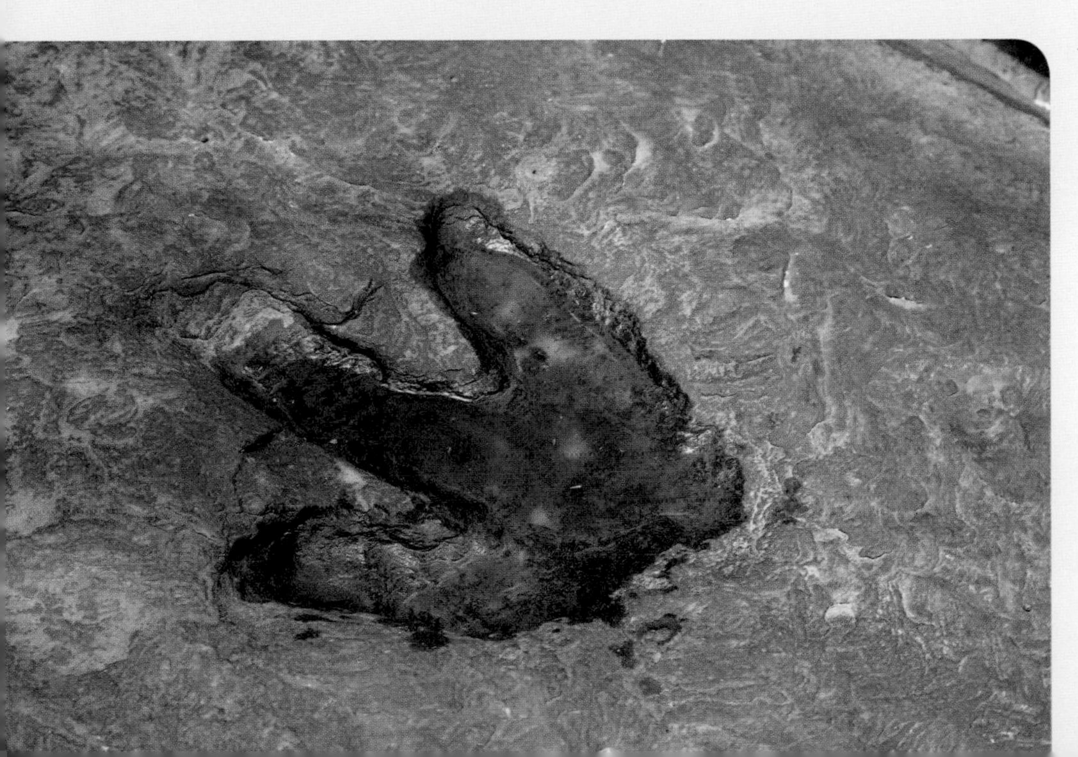

◀ As sediment gets compacted, it is slowly formed into sedimentary rocks. Any remains of animals and plants are fossilized. This photograph shows a fossilized dinosaur footprint. What else can rocks tell us about the Earth earlier in its history?

Key concept: Change

Related concepts: Transformation, Development

Global context: Fairness and development

Statement of inquiry:

The Earth has been transformed by slow processes that still continue today.

Introduction

The Earth is home to every living thing that we have encountered. Many scientists believe that it is only a matter of time before we find evidence of life, or its previous existence, elsewhere in the solar system or our galaxy. However, we currently know of no other place in the universe that has living organisms.

Because the Earth is unique in this way, and because it is the planet on which we live, scientists are keen to understand what the Earth is made of, and how the processes that shape the Earth work. In this chapter, we will look at the different rocks and other constituents that make up the Earth.

The Earth has existed for about 4.5 billion years and although it appears unchanging to us, it has slowly altered in its form and continues to do so. In this chapter, we will see how the Earth has changed over its history and how it is still changing. For this reason, the key concept of the chapter is change and the related concepts are transformation and development.

While the processes that slowly transform the Earth have been happening for billions of years, the actions of humans are also starting to affect the Earth. In this chapter, we will see how we have shaped the planet and how we might control these effects. As a result, the global context of the chapter is fairness and development.

Stone has been used for buildings and monuments for thousands of years. The Grand Menhir Brisé d'Er Grah (the Broken Menhir of Er Grah), in France, is one of the oldest in the world. It was erected in about 4700 BC and would have been about 20 m tall. It is thought to have fallen and broken around 4000 BC. What other stone buildings have survived for a long time?

The Grand Prismatic Spring is a hot spring in Yellowstone National Park, Wyoming. Its temperature in the center can reach 87°C but it is cooler towards the edges. Different bacteria thrive in different temperatures, which give the water different colors. The water in the pool permeates deep into the ground below, where it is heated by hot rock. This water then rises up into the spring, carrying the heat energy with it

What is the Earth made of?

When viewed from space, we can see that 71% of the Earth's surface is covered by water. At its deepest, the ocean is about 11 km deep. However, the average depth is about 3 to 4 km. Water is a hugely important part of the Earth. It is believed that life evolved in the oceans, and water is essential for all life on Earth.

While water is an important chemical and covers most of the surface of the Earth, it is not the most abundant substance on Earth. The most abundant elements in the surface layer of the Earth are oxygen, silicon and aluminium. As a result, chemical compounds such as silicon dioxide (SiO_2, also known as silica) and aluminium oxide (Al_2O_3, also called alumina) are the most common chemical compounds. Along with compounds of iron, calcium, sodium, potassium and magnesium, these make up the majority of the **minerals** on Earth.

These chemical compounds form small crystals in different structures, which are called minerals. For example, silicon dioxide forms a mineral called quartz and aluminium oxide forms the mineral corundum. It is possible for these chemicals to form different types of crystal. Calcium carbonate, for example, forms minerals called calcite, aragonite and vaterite. Many minerals have a more complicated chemical structure.

Rocks are made from small crystals or grains of mineral. Often these crystals are tiny, but sometimes large crystals such as gem stones can form. The type of rock that is formed will depend on the size of the crystals and the types of mineral that are present. For example, limestone rocks are formed from the minerals calcite and aragonite, while granite rocks are formed from quartz and a mineral called feldspar.

▲ Viewed from space, it is clear that the Earth's surface is covered in oceans

ABC

A **mineral** is a naturally occurring crystal form of a particular chemical substance.

A **rock** is a mixture of minerals bonded together into a solid formation.

▲ Sometimes, mineral crystals can grow quite large. This is a large (uncut) crystal of sapphire, which is a type of the mineral corundum. It contains impurities of titanium and iron, which give it its blue color. If the corundum had traces of chromium instead, it would have a red color and be called a ruby

Examining minerals

Use a magnifying glass to examine the surface of a variety of rocks. Some rocks have large grains. In other rocks, the grains are so fine that they cannot be seen.

ABC

Igneous rocks are formed from molten rock that cools and solidifies.

Sedimentary rocks are formed by material that is compacted together.

Metamorphic rocks are igneous or sedimentary rocks that are transformed by extreme pressure and temperature.

Rocks that are found on Earth fall into three categories: **igneous**, **sedimentary** and **metamorphic**.

What are igneous rocks?

Igneous rocks are the most common type of rock, although they are often found deep underground. They are formed when molten rock cools and solidifies. This might occur when a volcano erupts, in which case new igneous rock is formed from the eruption. However, some igneous rock is formed deeper in the Earth, where liquid rock cools and solidifies. Some of this type of rock will be deep under the Earth's surface and will be very old. Sometimes this old rock is exposed at the surface of the Earth.

Igneous rocks can be identified by looking for the following features:

- often very hard
- no layers
- mainly black, white or gray
- may have a flecked appearance.

▲ Pumice is an igneous rock that is formed in volcanic eruptions. When the molten lava cools, it solidifies, trapping bubbles of gas. This gives pumice a foam-like texture. The gas bubbles give pumice a low density and it can even float on water

▲ Granite is an igneous rock that is formed underground as magma cools. It is very hard and so does not get eroded quickly. This means that areas with granite can have high mountains and steep cliffs. This photograph shows El Capitan, a 900 m tall cliff face in Yosemite National Park, California. Glaciers shaped the rock over a million years ago. Because it is tough, granite is often carved into memorials such as gravestones

What are sedimentary rocks?

Sedimentary rocks are another type of rock. These are formed when small bits of other types of rock are eroded and are washed together. This material will be washed down rivers and will eventually be deposited, usually when the river runs into the sea. The material will include bits of rock, as well as mud, and may also include dead animals and plants. The material settles and is compressed by layers of sediment above it. As this material is compacted, the water is squeezed out and the material solidifies to form sedimentary rock. This process can take millions of years.

▲ Coal is a sedimentary rock that is formed from decayed plant matter. Coal can be burnt and used for fuel

The plant and animal matter that is included in the sedimentary rocks will also be transformed. As they are compacted, this matter will be converted into oil and coal over millions of years. Sometimes hard structures such as bones or shells can remain intact. These will form fossils.

Sedimentary rocks can be identified by looking for the following features:

- sometimes have layers
- may contain fossils
- may be possible to see grains of sand or sediment that were compacted
- they are usually brown or gray.

Because sedimentary rocks are formed near to the surface of the Earth, they are common. They account for about three quarters of the surface rock, but they do not exist deep underground.

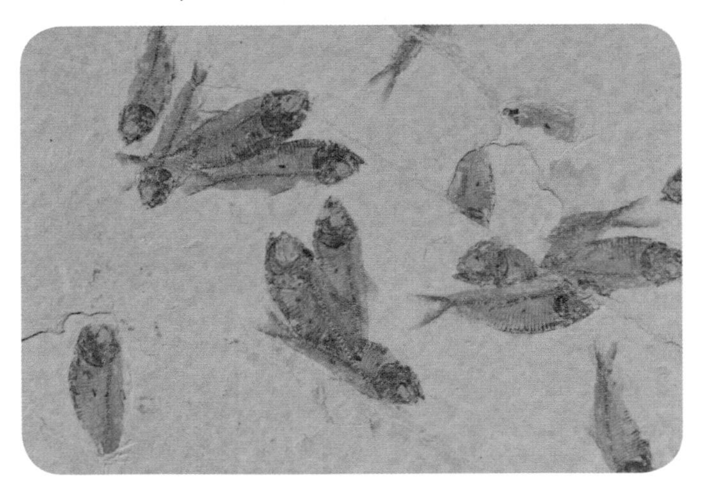

▲ Shale is a sedimentary rock formed by clay and mud that is washed down rivers and then compacted. Sedimentary rocks can have fossils in them. These fish fossils were created when the fish skeletons were trapped in the sediment and compacted

▲ Uluru, also known as Ayers Rock, is a large rock formation in Australia that is made of sandstone. Sandstone is a sedimentary rock that is created from compacted sand. Like many sedimentary rocks, it is relatively soft and therefore it has been eroded into smooth shapes

Limestone is a common sedimentary rock that is made of calcium carbonate. Because calcium carbonate reacts with acids and dissolves, acidic rain can erode limestone and create caves

What are metamorphic rocks?

Metamorphic rocks are rocks that have been transformed by intense heat and pressure. They are formed when existing igneous or sedimentary rocks are heated to temperatures of 150°C or more, and pressures that are 1,000 times the pressure in the atmosphere or greater. This alters the crystal structure of the minerals that make up the rock and often allows for large crystals to form.

The conditions to form metamorphic rocks may occur deep underground. Metamorphic rocks can also be formed closer to the surface of the Earth, where rocks come into contact with molten rock. This may happen near a volcano.

The high temperature and pressure cause the chemicals in the rocks to change. The crystals in the rock will reform and this will create a different texture to the original type of rock.

Metamorphic rocks can be identified by looking for the following features:

- may have different colors and colors may be bright
- may have wavy bands of color rather than layers.

Metamorphic rocks such as this lapis lazuli can have strong colors

1. Look at the rocks shown in images **a)** to **d)**. For each rock, try to identify whether it is igneous, sedimentary or metamorphic.

The Taj Mahal in India is made from white marble. Marble is a metamorphic rock that is formed when limestone is subjected to high temperatures and pressures. This causes the small calcite crystals in the limestone to reform into larger crystals, which interlock

How hard are minerals?

In Chapter 8, Properties of matter, we saw the Mohs scale, which is a measure of hardness. The Mohs scale was developed as a way of comparing the hardness of different minerals against each other. A hard mineral is able to scratch a softer mineral, but a soft mineral cannot scratch a harder one.

Ten minerals were used as reference points on the scale. These minerals and their hardness on the Mohs scale are shown in the table below.

▼ This picture shows the minerals quartz and calcite. One mineral is used to scratch the other. Which mineral is which?

Mineral	Hardness on the Mohs scale
diamond	10
corundum	9
topaz	8
quartz	7
feldspar	6
apatite	5
fluorite	4
calcite	3
gypsum	2
talc	1

1. Emery is a rock (sometimes called corundite) that is formed from the mineral corundum, while sand is usually formed from quartz. Both types of rock are used for polishing and smoothing, often by using sandpaper or emery paper. Which of these is more effective at polishing a hard surface?

What is inside the Earth?

ABC The **crust** is the outer layer of the Earth, consisting of solid rock.

The **mantle** is a layer of molten rock beneath the crust.

The **asthenosphere** is the top layer of the mantle, which behaves as a liquid.

The **core** is the central part of the Earth, consisting mainly of iron and nickel.

The rock that forms the surface of the Earth is called the **crust**. Under the oceans, the crust can be as little as 5 km thick. On land, the crust is about 35 km thick, and can be up to 70 km thick. This is far deeper than we can dig—the deepest mines are less than 4 km deep.

Beneath the crust lies a region of the Earth called the **mantle**. This consists of hot rock that can reach up to 4,000°C. Although the rock is very hot, the huge pressures inside the Earth keep the mantle in the solid state. Only the very top layer of the mantle—the **asthenosphere**—behaves as a liquid, although it is a very thick, sticky liquid that does not flow easily.

At the center of the Earth is the **core**. The core mainly consists of iron and nickel. The temperature is about 5,000°C, which is hot enough for the outer core to be liquid. The central part of the core, the inner core, has such high pressures that it is solid.

How does the Earth change?

The surface of the Earth changes so slowly that we would not expect to see any changes within our lifetimes. However, there is evidence that the continents have moved in the past and continue to do so today.

The shapes of the continents appear as if some of them would fit together, like a jigsaw puzzle. For example, the shape of South America appears to fit together with Africa very well.

Further evidence is found in the distribution of fossils. Fossils of a type of fern called *Glossopteris*, which lived more than 250 million years ago, are found in all of the southern continents, including Antarctica. It is unlikely that the fern's spores could have blown the vast distances across the oceans from Australia to India, Southern Africa and South America. It also suggests that the land mass of Antarctica was once situated somewhere warmer so that ferns could grow.

Today, we are able to measure the movement of continents despite their slow speed—they move only a few centimeters per year.

225 million years ago

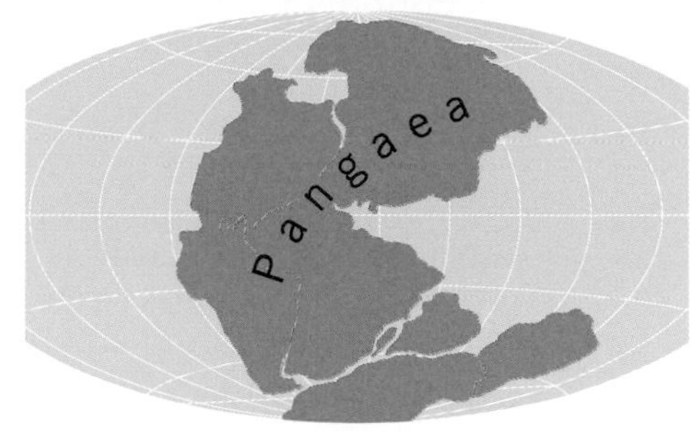

135 million years ago

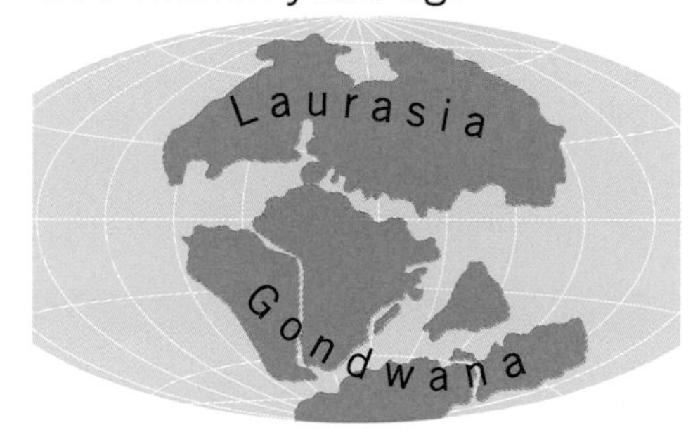

40 million years ago

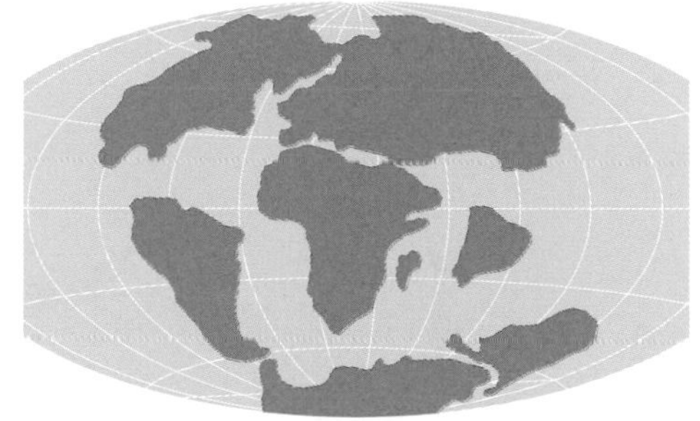

▲ Tectonic activity causes the continents to move slowly over time. At previous times in the Earth's history the land masses have been joined together to form a supercontinent. The last time this happened was over 200 million years ago when the land formed a supercontinent called Pangaea. This later broke into two landmasses called Laurasia and Gondwana. About 250 million years in the future, it is possible that the continents will again be joined together

▲ This is a fossil of a *Mesosaurus*, which was a type of reptile that lived in the water near the coast nearly 300 million years ago. *Mesosaurus* fossils are found in the east of South America and Southern Africa. Because these animals would not have been able to swim across the Atlantic Ocean, this provides evidence that South America and Southern Africa were once much nearer together

1. Fossils of *Mesosaurus* are found in South America as well as in Southern Africa. These two locations are now 6,500 km apart. The fossils suggest that the two continents moved apart about 300 million years ago. Calculate the speed at which the continents are moving apart.

ABC **Continental drift** is the slow motion of the Earth's continents.

A **tectonic plate** is a region of the Earth's crust that moves as a whole.

A **tectonic boundary** is the region between two tectonic plates. Tectonic boundaries may be **convergent** when the plates are moving together, **divergent** when they move apart, or when the plates slide past each other they form a **transform plate boundary**.

How do the continents move?

The motion of the continents is sometimes called **continental drift**, but the theory of how this happens is called plate tectonics. The Earth's surface can be divided into large regions that are called **tectonic plates**. Each plate moves as a solid object, which can move across the more liquid rock in the asthenosphere below.

In between these tectonic plates is a **tectonic boundary**. These boundaries will be where the tectonic plates are moving apart, moving together or sliding past each other.

A boundary where the tectonic plates are moving away from each other is called a **divergent** boundary. As the plates move apart, molten rock from the mantle underneath rises up and cools, forming new igneous rocks. The Mid-Atlantic Ridge is a good example of a divergent boundary. In the middle of the Atlantic Ocean, the North American plate moves westward, and the Eurasian Plate moves eastward. Further south, the ridge separates the South American Plate and the African Plate.

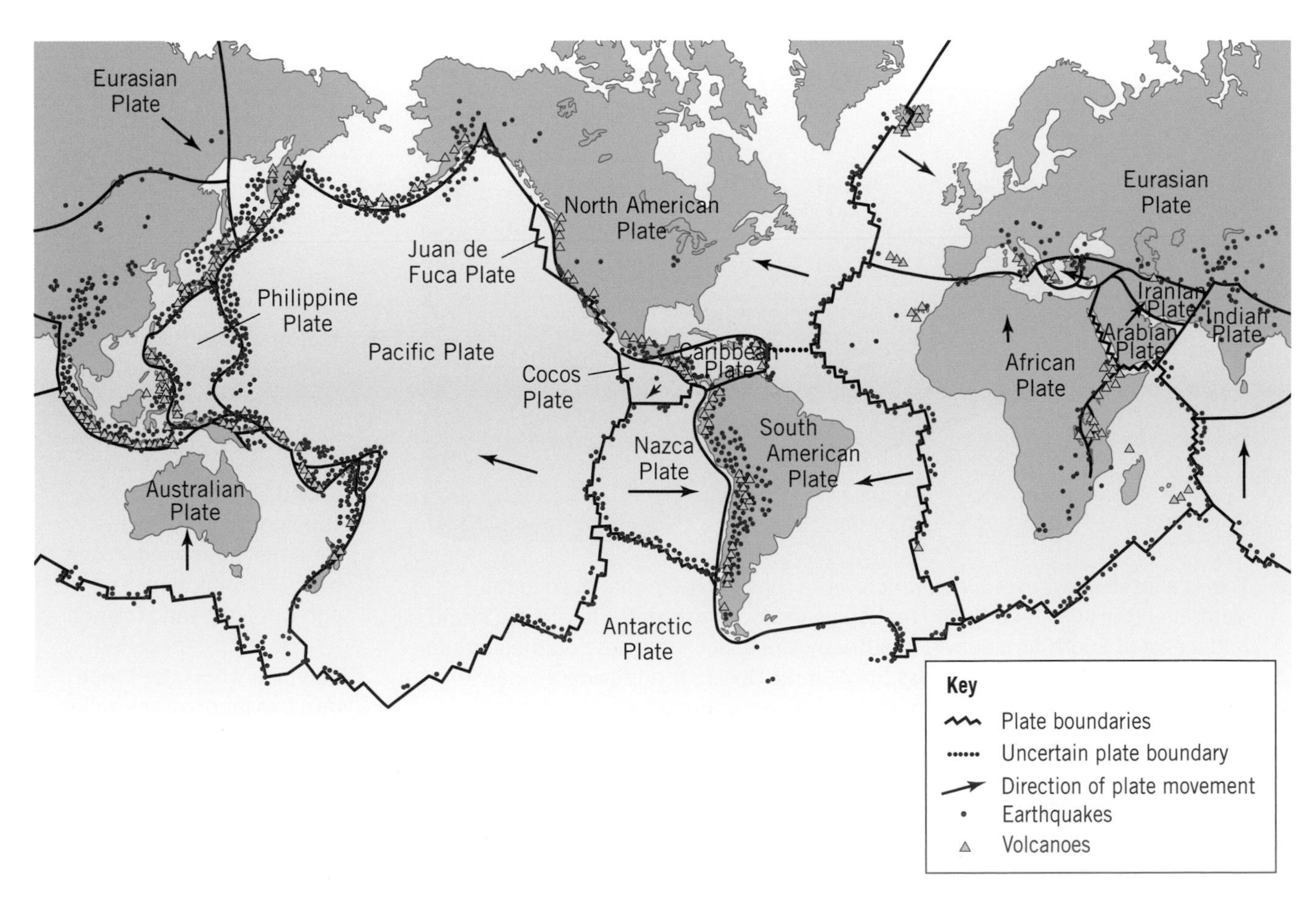

▲ This diagram shows the major tectonic plates of the Earth and the direction in which they are traveling. The large number of volcanoes that surround the Pacific Ocean are called the ring of fire

◀ The Azores islands lie on the Mid-Atlantic Ridge on the boundary between the North American Plate, the Eurasian Plate and the African plate. They are formed from volcanic activity

▲ This photograph was taken after the 1906 San Francisco earthquake at the San Andreas Fault. Before the earthquake, the fence would have linked together in a straight line. During the earthquake, the two tectonic plates moved past each other, separating the parts of the fence at the boundary

Another type of tectonic boundary occurs when two plates slide past each other. This type of boundary is called a **transform plate boundary**. An example of this type of tectonic boundary is the San Andreas Fault in California, where the North American Plate slides past the Pacific Plate. The two plates at a transform boundary do not move past each other smoothly. Instead, the forces build up until the plates move suddenly, causing an earthquake.

Other tectonic boundaries occur where two plates are moving together. These are called **convergent** boundaries. When this happens one plate is pushed under another. This process is called subduction and the area where this happens is called a **subduction zone**. When a tectonic plate is subducted, deep trenches in the ocean can be formed. An example of an **oceanic trench** is where the Pacific Plate is subducted under a small tectonic plate called the Mariana Plate, near the Philippine Plate. The Mariana Trench was formed as a result. It contains the deepest known point in the ocean, known as the Challenger Deep, which has a depth of nearly 11 km.

Convergent boundaries surround the Pacific Ocean. This has caused there to be a large number of volcanoes around Indonesia, Japan and the Philippines, and the west side of North and South America. This ring of volcanoes is often called the ring of fire.

ABC

A **subduction zone** is a convergent tectonic boundary where one tectonic plate is pushed underneath another.

An **oceanic trench** is a deep trench in the ocean near a convergent tectonic boundary.

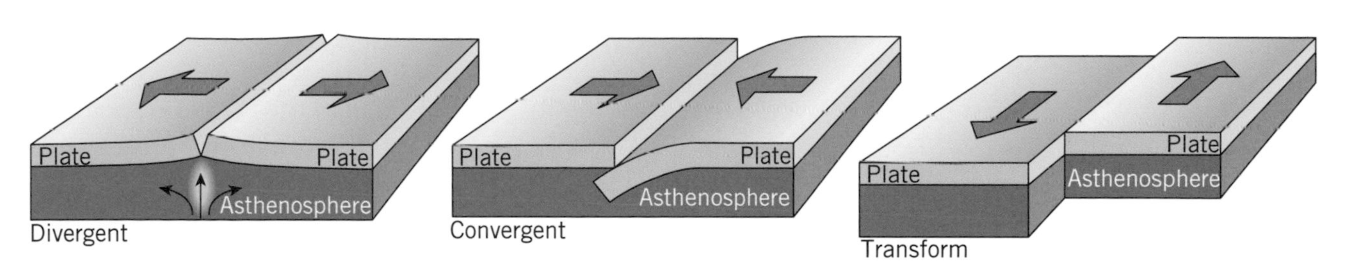

▲ Different types of tectonic plate movement

While one plate sinks and is subducted, the other tectonic plate is pushed upwards. The rock that forms the floor of the ocean is usually denser than the rock that forms the land masses. As a result, the subducted plate is usually the one under the ocean, and the plate that is pushed upwards usually has the land on it. As the land is pushed upwards, mountain ranges can be formed.

▲ The Himalayas contain the highest mountains on Earth. The mountain range was created as the Indian Plate was subducted underneath the Eurasian Plate, which pushed the land upward, creating the mountain range. This process is still happening, and the Himalayan mountains get taller by about 5 mm per year

Subduction zones are also associated with earthquakes and volcanoes. As one tectonic plate is pushed under the other, the forces acting on the plates is enormous. The two plates do not slide freely but instead get stuck, and then suddenly move. This causes large earthquakes when the plates move. The tectonic plate that is subducted will be pushed down towards the mantle. As this happens, the rock in the mantle is melted and gases such as carbon dioxide can be released. As these gases rise, they heat up the rock above them, melting it and forming magma. This results in the formation of a volcano.

1. The Pacific Plate moves at about 70 mm per year from a divergent plate boundary on one side to a subduction zone on the other side. These two boundaries are approximately 10,000 km apart. How long would it take for the rock on one side of the Pacific Plate to move from the divergent boundary to the subduction zone?

Tectonic plates

The table below shows the areas of the ten largest tectonic plates.

There are many other, smaller tectonic plates.

1. The total surface area of the Earth is 510 million km^2. What is the total area of all the other, smaller tectonic plates?
2. What fraction of the Earth's surface area is covered by the Pacific Plate?
3. Draw a pie chart to show the areas of the tectonic plates as fractions of the Earth's total surface area.

Tectonic plate	Area (million km^2)
Pacific	103.3
North American	75.9
Eurasian	67.8
African	61.3
Antarctic	60.9
Australian	47.0
South American	43.6
Somali	16.7
Nazca	15.6
Indian	11.9

What is a volcano?

A volcano is a place where molten rock and ash reach the surface of the Earth. Molten rock in the Earth's crust is called **magma** but when it reaches the surface of the Earth it is called **lava**. When they erupt, they are one of the greatest natural hazards, as they are capable of producing huge explosions. Volcanoes are mostly found near divergent and convergent tectonic boundaries.

ABC

Magma is molten rock in the Earth's crust.

Lava is molten rock on the Earth's surface.

Volcanoes that are currently erupting or that have regular earthquakes are called active volcanoes. These volcanoes are not stable. Other volcanoes are not currently erupting but could erupt in the future; these are called dormant volcanoes. Some volcanoes have not erupted for millions of years and are thought to be inactive and incapable of erupting again. Such volcanoes are called extinct volcanoes. Distinguishing between these types of volcano can be difficult as a volcano can be dormant between eruptions for many thousands of years, or even longer.

Some volcanoes occur at divergent tectonic boundaries. Here the magma rises up as the plate boundaries move apart to fill the gap. The magma does not reach high pressures and has a chemical constituency that allows it to flow. As a result, the lava that reaches the surface often flows gently out of the volcano.

Other volcanoes occur at convergent boundaries. Here, the subducted rock is melted as it reaches the mantle, releasing water and gases from the rock. This can cause the magma to have a higher pressure. The constituents of the magma are also different, causing it to be more sticky and viscous. When this type of volcano erupts, it will do so in a massive explosion.

▲ This photograph shows lava flowing from a volcano in Iceland. Iceland sits on the Mid-Atlantic Ridge, which is a divergent tectonic boundary. This means that Iceland has a lot of volcanic activity. However, as the volcanoes are at a divergent boundary, the lava can flow easily. This means that the pressure inside the volcanoes stays low and the eruptions are not explosive

ABC The **Richter scale** is a measure of the size of an earthquake.

A **seismograph** is an instrument used for measuring earthquakes and tremors.

A **tremor** is a very small earthquake.

How can we measure earthquakes?

An earthquake occurs when the ground shakes. This is usually caused by the movement of tectonic plates. Earthquakes are therefore more common and stronger near tectonic boundaries, although weak earthquakes can occur anywhere in the world.

The size or magnitude of an earthquake can be measured on the **Richter scale**. This is a logarithmic scale: an increase of 1 on the Richter scale means that the amplitude of the seismic waves is ten times greater. The device used to measure an earthquake is a **seismograph**. This device can measure tiny **tremors** in the ground.

If an earthquake has a magnitude less than 2, it might be detected by seismographs, but it is unlikely any people would feel it. This type of earthquake happens all the time. If an earthquake has a magnitude of 4, then it is strong enough to shake objects, but it is unlikely to cause much damage. At a magnitude of 5, an earthquake will start causing damage. If an earthquake measures 6 on the Richter scale, it will likely cause damage to buildings. About 100 earthquakes per year exceed 6 on the Richter scale.

About ten earthquakes per year are major earthquakes, registering 7 or more on the Richter scale. These are felt around a large area and can cause a lot of damage. An earthquake that has a magnitude of 8 or more would be very destructive to a large area. These type of earthquakes only occur approximately once per year.

Occasionally, an earthquake is large enough to exceed a magnitude of 9 on the Richter scale. These earthquakes cause widespread destruction but are rare. The most powerful earthquake ever recorded had a magnitude between 9.4 and 9.6 on the Richter scale. This earthquake occurred in Chile in 1960.

This photograph was taken after the 2015 earthquake in Nepal, which had a magnitude of 8.1

Earthquake magnitudes

This graph shows the energy that is thought to be released from earthquakes that have a magnitude between 7 and 8 on the Richter scale.

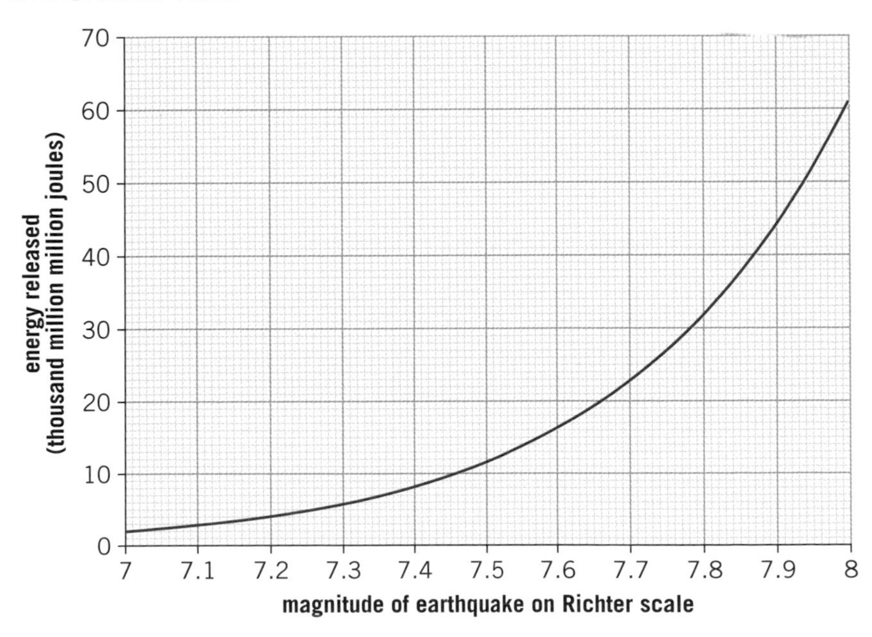

1. On 22 February 2019, an earthquake of magnitude 7.5 struck in Ecuador. Estimate the amount of energy that was released.
2. How many times greater is the energy released from an earthquake of magnitude 8.0 than from an earthquake of magnitude 7.0?

ATL **Creativity and innovation skills**

Earthquakes in cities

Some cities such as Tokyo and Los Angeles are located near tectonic boundaries. As a result, they experience many earthquakes. Tokyo experiences a detectable earthquake almost every week although the majority of these are weak.

As a result of the frequent earthquakes in Japan, all new buildings have to be designed to withstand strong earthquakes. These innovative buildings have foundations that absorb the energy of the earthquake. They are also built to be flexible so that they are not damaged when they shake.

▲ The Roppongi Hills Mori Tower is one of Tokyo's tallest buildings. The foundations are built with oil-filled shock absorbers that control the swaying of the building in an earthquake

Are other planets like the Earth?

▲ Maat Mons, the highest volcano on Venus, is 8 km high. It is unknown whether it is active, or whether it has been dormant or even extinct for a long time

Studying other planets enables us to draw comparisons with the Earth. Scientists want to know whether the same processes that occur on Earth are also happening on other planets—or is the Earth unique?

The planets that are most like Earth are Venus and Mars. Venus is often regarded as Earth's sister planet because it is a similar distance from the Sun and is only a little smaller than the Earth. However, its dense atmosphere creates highly inhospitable conditions on the surface. The temperature at the surface is about 460°C and the pressure is almost 100 times the atmospheric pressure on Earth. The longest time that a probe has managed to operate on the surface is just over two hours, when the *Venera 13* mission landed in 1982. Despite this, scientists have been able to use radar measurements to image the surface of Venus and have found thousands of volcanoes. However, none have been seen to erupt and so it is not known whether Venus is still volcanically active. It is not thought that Venus has any tectonic activity.

Mars, while further from the Earth, is easier to study. It also has volcanoes, the tallest being Olympus Mons, which is nearly 22 km high. It is possible that Mars once had tectonic activity, although this has not happened for many billions of years. While most of the rocks are igneous, sedimentary rocks have also been found. This indicates that Mars may once have had liquid water flowing over its surface.

Mars has many extinct volcanoes. Olympus Mons, the largest, is seen to the bottom right of the picture and is almost 22 km high

While Mars and Venus have extinct volcanoes, there are some places in the solar system where active volcanoes have been found. These volcanoes eject water and ice instead of hot rock. They are called **cryovolcanoes**. Cryovolcanoes have been found on some moons of Jupiter, Saturn and Neptune and it is likely that they exist elsewhere in the solar system as well. To fuel these volcanoes, the moons need a source of heat. As the moons orbit these giant planets, the large gravitational forces cause the moon to be flexed and bent as it rotates. The frictional forces that act against the deforming of the moon transfer energy to heat and heat up the inside of the moon. It is thought that the insides of these moons have liquid water in the form of giant underground oceans.

1. Why do astronomers get excited about discovering liquid water on other moons or planets?

ABC A **cryovolcano** is a volcano that has eruptions of ice and water. They occur on moons and planets in the solar system.

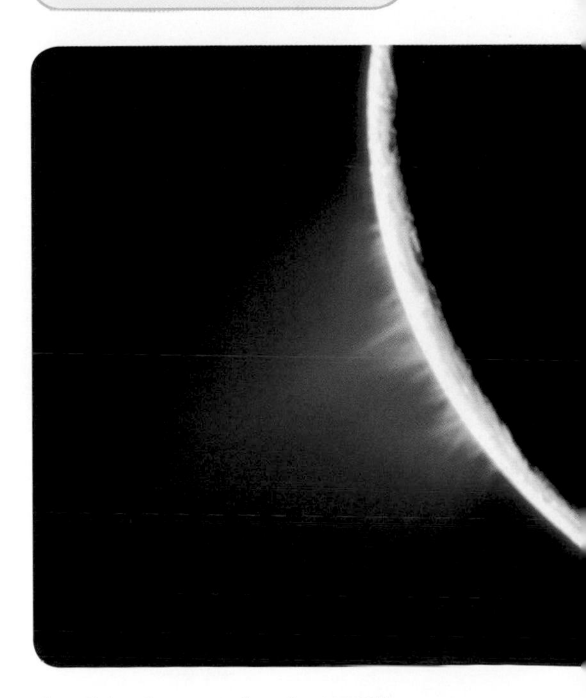

This picture, taken by NASA's *Cassini* probe, shows plumes of ice that are ejected from Enceladus, a moon of Saturn

How have humans changed the Earth?

It is tempting to see the Earth as being too large to be changed by human activity. After all, the total mass of all humans (about 385 million tonnes) is much less than the huge mass of the Earth (about 6000 million million million tonnes). However, the action of humans is significant, and some scientists have proposed that the Earth has entered a new age—the Anthropocene—whereby human activity is the dominant effect on the world.

One way in which human activity has affected the world is through nuclear tests. The first nuclear detonation in 1945 created radioactive isotopes that still exist today. In future ages, sedimentary rocks will show a layer where there is a higher amount of these isotopes (see Chapter 7, Atoms, elements and compounds, for more on isotopes).

▲ The Trinity test took place on 16 July 1945 and was the first nuclear device to be detonated

Another new layer that will occur in sedimentary rocks is from our use of plastics. We produce hundreds of millions of tonnes of plastic every year. Some of the plastic ends up in landfill, but some of it breaks down into small particles and these can find their way into the oceans. Small particles of plastic can join the sediment and start to become part of the sedimentary rocks that form. This layer may be detectable millions of years in the future.

Summative assessment

Statement of inquiry:

The Earth has been transformed by slow processes that still continue today.

In August 1883, the volcanic island of Krakatoa in Indonesia erupted. The explosion was so violent that it destroyed the majority of the island. The sound was so loud that it was heard 4,780 km away near Mauritius. This assessment is about the Krakatoa volcano.

▲ This photograph shows an eruption of Anak Krakatau (Child of Krakatoa) in Indonesia. In 1883, this was the site of one of the largest volcanic eruptions ever recorded, which caused 36,000 fatalities and destroyed a large part of the island. Anak Krakatau appeared above the water in 1927 and has grown to a height of over 100 m. It regularly erupts—in December 2018 it erupted causing a deadly tsunami

Volcanoes and tectonic plates

1. The island of Krakatoa sits on a site where two tectonic plates are moving toward each other. Which name is given to this kind of boundary?

   A. Convergent boundary

   B. Divergent boundary

   C. Transform plate boundary

   D. Volcanic boundary

2. Which of these features would you not expect to find near a tectonic boundary where two plates are moving toward each other?

   A. Subduction zone **C.** Oceanic ridge

   B. Oceanic trench **D.** Volcano

3. The two tectonic plates near the island of Krakatoa are moving toward each other at a speed of 0.16 mm per day. How far do they move in one year?

   A. 0.6 cm **B.** 5 cm **C.** 6 cm **D.** 58 cm

4. The sound from the volcano was heard 4,780 km away. If the speed of sound is 330 m/s, how long would it have taken the sound to travel this distance?

   A. 240 seconds

   B. 4 minutes

   C. 4 hours

   D. 240 hours

5. A volcano is caused by a build-up of molten rock. What name is given to this molten rock when it is under the ground?

   A. Lava

   B. Magma

   C. Igneous

   D. Granite

6. The explosion at Krakatoa is estimated to have released the energy equivalent to 200 million tonnes of TNT. The first nuclear detonation—the Trinity test—yielded an equivalent of 22 thousand tonnes of TNT. How many times more powerful was the explosion at Krakatoa?

   A. 9

   B. 900

   C. 9,000

   D. 90,000

The stickiness of lava

The chemical constituents of the lava that flows out of a volcano affect its viscosity. Viscosity is a measure of how resistant a fluid is to flowing, or how well it sticks together. For example, honey is more viscous than water because it doesn't flow as easily. Explosive volcanoes such as Krakatoa have molten rock with a high viscosity, which means the rock does not flow well.

Volcanologists (scientists who study volcanoes) believe that a higher proportion of silica (SiO_2) gives the lava a higher viscosity. Other factors, such as temperature, can also affect the viscosity of lava. To test this theory, they must study the chemical composition of volcanoes and how the lava flows out of them.

▲ Lava is usually erupted at a temperature between 700 and 1,200°C

7. Give one difficulty associated with measuring the viscosity of lava. [2]
8. A volcanologist decides to test the hypothesis that a higher content of silica will result in more viscous lava.
   a) What is the independent variable for this investigation? [1]
   b) What is the dependent variable for this investigation? [1]
   c) Suggest a control variable for the investigation. [1]
9. Volcanologists often use satellite data when monitoring volcanoes. Suggest why this might be a good idea. [2]
10. Explain why the volcanologist would need to collect data from many volcanoes around the world, and for a long time, in order to test the hypothesis. [3]

Anak Krakatau

In 1927, Anak Krakatau (Child of Krakatoa) first appeared above the ocean. Since then, it has grown in size. A table of the height of the island as reported in different years is shown in the table below.

Year	1952	1957	1960	1977	1982	1987	1992	2006	2011	2018	2019
Height (m)	70	170	138	181	200	194	250	300	324	338	110

11. The heights in the table were reported from different sources. Explain one problem with taking data from different sources. [2]

12. Plot a graph of the data in the table. [4]

13. Calculate how much height the volcano grew in an average year between 1960 and 2018. [1]

14. An eruption in December 2018 significantly reduced the height of the volcano. Using your answer to the previous question, how long might it take for the volcano to grow back to its height before the 2018 eruption? [2]

15. It is thought that the height of the Krakatoa volcano before it exploded in 1883 was 813 m.

a) How long would it take for the volcano to reach this height again? [1]

b) Explain one assumption made in your calculation. [1]

Studying volcanoes

Scientists study Krakatoa and other volcanoes in an attempt to predict when they might explode. One way of doing this is to monitor seismic activity at the volcano.

16. Describe the benefits of this research. [2]

17. Describe what problem volcanoes present. [3]

18. Some scientists have suggested that sites where tectonic plates converge are suitable places for disposing of hazardous nuclear waste. The idea is that the waste would be taken towards the center of the Earth over time.

a) Describe the problem that is solved by this idea. [2]

b) Suggest one problem with this idea. [2]

11 The atmosphere

Key concept: Systems

Related concepts: Consequences, Environment

Global context: Globalization and sustainability

◀ Water vapor in the atmosphere forms clouds and eventually falls as rain. Rain is important for life. It brings water to the land, dissolves soluble gases from the atmosphere and washes out dust. Which other processes make life possible?

▶ The 1968 Olympic Games were held in Mexico City, which is at an altitude of 2,200 m. The lower atmospheric pressure helped athletes to produce fast times in speed events. The world records for the men's and women's 100 m and 200 m were broken, and it was the first time that the 100 m was run in under 10 s, and the 200 m in under 20 s. Another world record was set in the men's long jump when Bob Beamon jumped 55 cm further than anyone had before, with a jump of 8.9 m. This world record stood for 23 years and is still the second furthest jump of all time. How else can the environment affect our performance?

Statement of inquiry:

The atmosphere around us creates the conditions necessary for life.

▲ The Earth isn't the only object in the solar system to have an atmosphere. There are many planets, dwarf planets and moons that have an atmosphere. In 2015, the *New Horizons* probe passed Pluto and observed its atmosphere. The atmospheric pressure at the surface of Pluto is about 100,000 times less than on Earth and therefore the atmosphere is much thinner. However, it still creates a greenhouse effect and there may even be clouds. Why do you think that scientists are interested in atmospheres on planets other than the Earth?

◀ Humans pollute the atmosphere with emissions from vehicles and factories. This can become a serious problem in large cities. What are the impacts of this pollution?

Key concept: Systems

Related concepts: Consequences, Environment

Global context: Globalization and sustainability

Statement of inquiry:

The atmosphere around us creates the conditions necessary for life.

Introduction

The Earth's atmosphere is vital for supporting life. It contains the air that we breathe, it is responsible for the weather that brings water onto the land and it protects us from some of the Sun's more harmful radiation. Without the atmosphere, there would be no possibility of life on the Earth. In this chapter, we will investigate what the atmosphere is and what it does. We will look at some of the systems such as the water cycle and weather systems that occur in the atmosphere. For this reason, the key concept of the chapter is systems. One of the related concepts is environment because of the way in which weather and the atmosphere affect the local environment.

Human activity is changing the nature of the atmosphere and the consequences of this are still yet to be fully understood. As we release quantities of greenhouse gases into the atmosphere, we add to the greenhouse effect and alter some of the balanced environmental systems. The second related concept of the chapter is consequences and the global context is globalization and sustainability.

▶ The Earth's atmosphere has been different in the past. It is thought that between 350 and 300 million years ago, in the Carboniferous period, the amount of oxygen in the atmosphere was significantly higher than it is today. This enabled the insects that inhabited the Earth at the time to evolve into large creatures. The fossil in this picture is from a millipede, *Arthropleura*, and only contains part of its leg. The fossil itself is about 7 cm in size; it is thought that these millipedes could be 50 cm wide and over 2 m long

▶ As the Sun's light passes through the atmosphere, the rays of light are scattered by dust in the air and the air molecules themselves. Blue light is scattered more, and this scattered blue light will therefore arrive at all angles and the sky appears to be blue. When the Sun is low in the sky, at sunset for instance, the light passes through more of the atmosphere. The blue light is scattered away, leaving more red light

What is our atmosphere made of?

The atmosphere is mainly formed of nitrogen gas (N_2), which accounts for 78% of the atmosphere. The next most common constituent is oxygen gas (O_2), which accounts for 21%. Together, these two gases account for 99% of the atmosphere. After these, the next most common constituents are water vapor, argon and carbon dioxide.

The nitrogen in the air does not react easily. However, nitrogen compounds form important nutrients for plant growth and are important for animals to make proteins. Some plants such as peas and beans, as well as some bacteria, can convert nitrogen from the air into soluble nitrogen compounds. Lightning can also convert nitrogen into chemical compounds that can be used by plants and animals. Chemical compounds are discussed in Chapter 7, Atoms, elements and compounds.

The oxygen in the air is essential for the process of respiration. This process allows animals to breathe and they use the oxygen in chemical reactions that release energy. As they do so, they convert oxygen into carbon dioxide. The process of respiration occurs in all organisms.

The oxygen is replaced as plants use the Sun's energy to convert carbon dioxide back to oxygen. This process is called photosynthesis.

▼ This diagram shows the oxygen cycle. All organisms require oxygen for the process of respiration. Combustion also requires oxygen. The oxygen in the atmosphere is replaced by the process of photosynthesis in plants

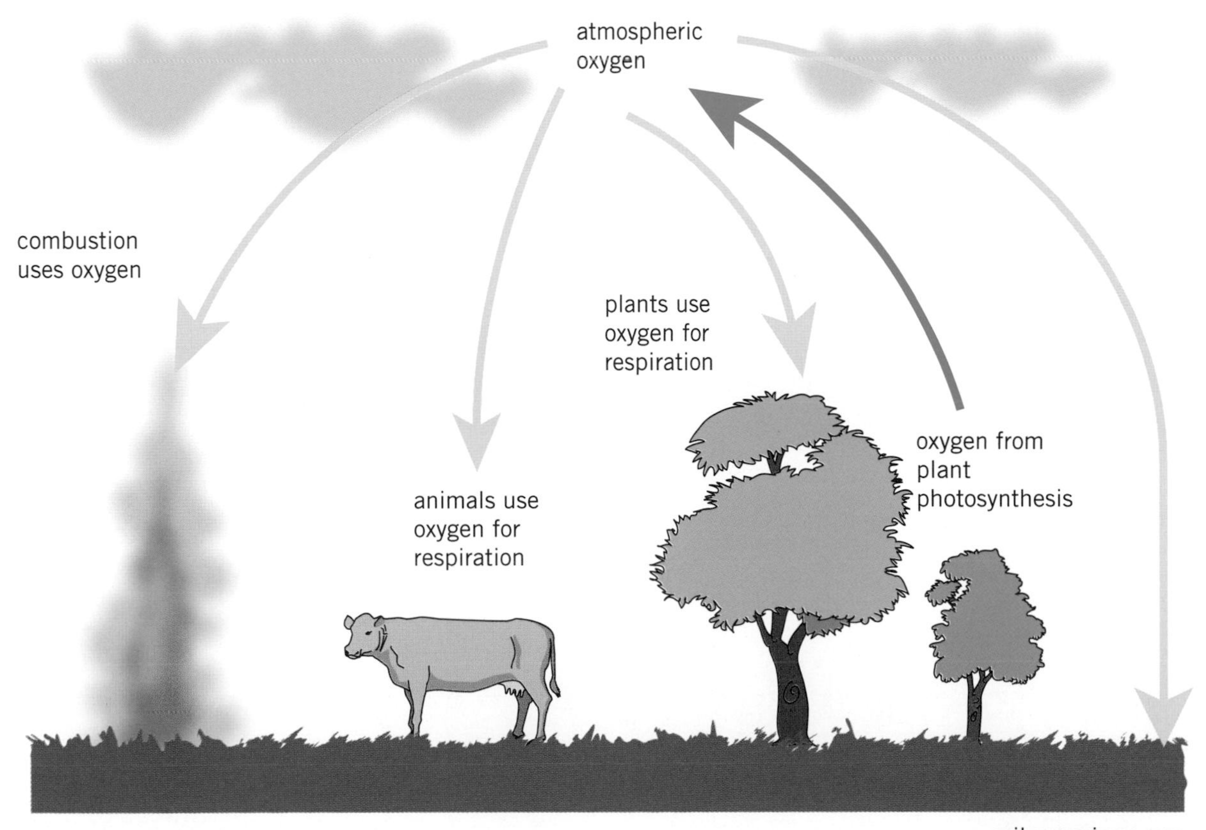

ABC

Altitude is the height above the surface of the Earth.

The **troposphere** is the lowest layer of the Earth's atmosphere.

The **stratosphere** is the second layer of the atmosphere above the Earth's surface. It contains the ozone layer.

The **ozone layer** is a region of ozone gas in the stratosphere. It absorbs harmful UV radiation from the Sun.

Constructing pie charts

When a set of data gives the components that make up a whole, it is often useful to present the data as a pie chart. Use the table below to construct a pie chart that shows the constituents of the atmosphere. To do this you will need to calculate the angle that each slice of the chart should occupy.

Constituent	Proportion of the atmosphere (%)	Angle represented on the pie chart (°)
nitrogen	78	281
oxygen	21	
other	1	

How does the atmosphere change with height?

If you were to travel upwards, the atmosphere would change. The air pressure, density and temperature all change with increasing **altitude**.

The atmosphere is not uniform; instead, it is divided into layers. The lowest layer is the **troposphere**, which spans from the ground up to about 12 km in height. The troposphere is the densest layer of the atmosphere and most of the mass of the atmosphere is in this layer. The clouds are within the troposphere and most planes will also fly in this layer of the atmosphere. Even the top of Mount Everest is still within the troposphere.

As you move upwards through the troposphere, the temperature and pressure decrease. At the top of the troposphere, temperatures are about –80°C, and the pressure is a tenth of what it was at sea level.

▼ The atmosphere at the top of Mount Kilimanjaro (in the background of this picture) is much thinner and colder than at the base of the mountain. The air pressure at the top is about half that at sea level, which makes climbing Mount Kilimanjaro difficult

Above the troposphere is a layer called the **stratosphere**. The stratosphere extends from the top of the troposphere to about 50 km above the Earth. In the stratosphere, a type of oxygen gas called ozone (O_3) is found. Ozone absorbs harmful ultraviolet rays from the Sun. This layer of ozone, referred to as the **ozone layer**, helps to shield the surface of the Earth from these ultraviolet rays.

The atmospheric pressure in the stratosphere is 10 to 1,000 times lower than at sea level.

Temperatures in the stratosphere actually increase with height because of the ultraviolet light that is absorbed.

Beyond the stratosphere is the **mesosphere**, which reaches up to 90 km above the surface of the Earth. The mesosphere has a low pressure (1,000th to 100,000th of sea level air pressure) and the temperatures decrease from around −15 °C to about −150 °C.

The mesosphere is the layer of the atmosphere that is responsible for shooting stars. Shooting stars, or **meteors**, occur when a small rock, a small asteroid or a fragment of a comet falls into the Earth's atmosphere from space. As they fall, they get heated by air resistance. A small rock will get so hot that it completely burns up in the mesosphere. Sometimes, if the rock is large enough, fragments will reach the ground. This is a **meteorite**.

ABC The **mesosphere** is the layer of the atmosphere between 50 and 90 km above the surface of the Earth.

A **meteor** is a small fragment of rock that falls into the Earth's atmosphere and burns up in the atmosphere.

A **meteorite** is a rock from space that enters the Earth's atmosphere and falls to the ground.

Meteor showers are caused by rocks from space burning up in the mesosphere

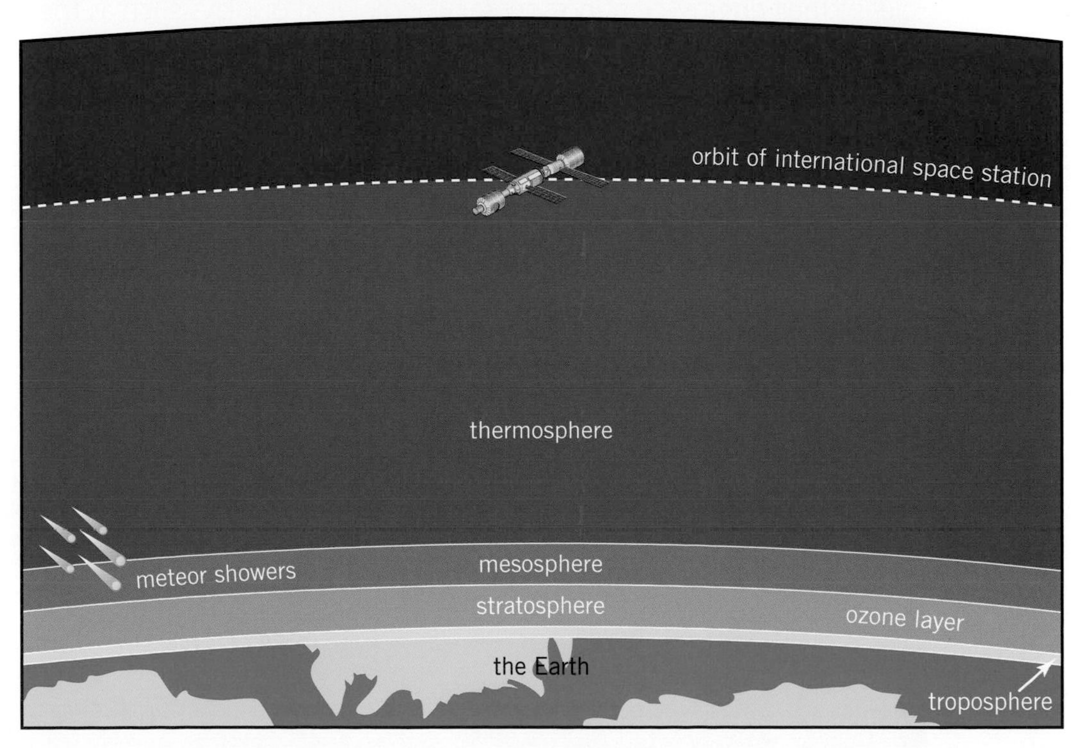

This diagram shows the layers of the Earth's atmosphere. The diagram is to scale and shows the Earth's curvature. The troposphere, which contains the entire breathable atmosphere, is a tiny layer when compared to the size of the Earth

ABC The **thermosphere** is the outer layer of the Earth's atmosphere, between 90 and 600 km above the Earth. The thermosphere is considered to be the start of space.

Above the mesosphere is the **thermosphere**, which extends to about 600 km above the surface of the Earth. At this height, the atmospheric pressure is so low that satellites can orbit at this altitude without experiencing significant air resistance. The International Space Station (ISS) orbits in the thermosphere and the air resistance from the thermosphere only causes it to lose about 60 m of altitude every day. The ISS, and other satellites that orbit in the thermosphere, need a periodic rocket thrust to keep them in orbit.

▶ The International Space Station (ISS) orbits at a height of about 400 km. Although this is inside a layer of the Earth's atmosphere called the thermosphere, the density of the air is so low (approximately 2 million million times lower than at sea level) that the space station and other satellites can orbit freely

Auroras occur in the thermosphere. Charged particles from the Sun are guided by the Earth's magnetic field towards the poles. Near the North and South Poles, the charged particles enter the thermosphere and release their energy, creating spectacular glowing displays such as the northern lights, or aurora borealis.

▶ An aurora is caused by charged particles from the Sun hitting the thermosphere and losing energy

1. When a small rock enters the atmosphere, it burns up in the mesosphere as a meteor. Describe the energy change that occurs as the rock burns up.
2. Outer space is defined for legal purposes as starting at 100 km. In which part of the atmosphere is this?
3. Planes on long journeys sometimes fly at the base of the stratosphere. They do not fly higher than this because they remove ozone from the stratosphere. Give one advantage and one disadvantage of flying at a high altitude.

How does our atmosphere enable life?

The atmosphere is essential for providing an environment in which life can exist. One of the ways it does this is by enabling liquid water to exist.

If you were to climb a mountain, the atmospheric pressure would decrease. As it did so, you would find that the boiling point of water also decreased. At an altitude of 1,000 m, the decreased pressure would mean that water would boil at about 96°C rather than 100°C. At the top of Mt Everest, the lower pressure would make water boil at 72°C. At an altitude of 18 to 19 km, the boiling point of water gets lower than body temperature. This is called the Armstrong limit: to survive at this altitude and low pressure requires a pressurized space suit. If you kept going upwards, the pressure would drop so low that the boiling point of water would become equal to the melting point. As a result, ice would sublimate directly to water vapor and liquid water could not exist. (See Chapter 8, Properties of matter, for more on sublimation and other changes of state)

▲ Astronauts in space need to wear a pressurized space suit so that the boiling point of water doesn't become too close to body temperature

Scientists think that liquid water is essential for life. Many chemicals can be dissolved in water, and this allows them to flow easily around an organism. Mammals have a blood supply that uses liquids to transport nutrients and oxygen, and plants have xylem and phloem, which achieve the same result. The xylem moves water and minerals from the soil that are absorbed by the roots up the stem and to the leaves of the plant. The phloem moves the chemicals that the plants create in their leaves around the plant. These substances are an energy store, similar to food, for the plant and might be used for the plant to grow, or might be stored.

The liquid involved in life does not necessarily have to be water—it is possible that life could form on other planets using a different liquid substance. However, water is a compound consisting of hydrogen and oxygen atoms. These are common elements and therefore water is often thought to be the most likely liquid to support life. Since liquids require an atmospheric pressure to exist, it is thought that life would evolve on planets that have an atmosphere.

We have also seen that the stratosphere contains the ozone layer, which shields the Earth from harmful UV rays. Without this ozone layer, the high energy UV rays would disrupt the chemistry that enables life. Life could exist underground or underwater, but most of the life on Earth would not be possible without the ozone layer.

What is the greenhouse effect?

The greenhouse effect occurs in the atmosphere. This is another way that the atmosphere creates the conditions on the Earth's surface that enable life to flourish.

The Sun's heat and light warm the surface of the Earth. Because the Earth's surface is warm, it too radiates heat energy in the form of infrared waves. However, because the Earth is much cooler than the Sun, the wavelength of these waves is longer.

The atmosphere contains water vapor and carbon dioxide, which absorb the longer wavelength infrared waves. The absorbed radiation heats up the atmosphere and this warm air radiates heat energy back to the Earth's surface. As a result, the atmosphere acts like a layer of insulation, keeping the Earth's surface warm. This is called the greenhouse effect.

Without the greenhouse effect, the surface of the Earth would be frozen. Instead, the average surface temperature of the Earth is about 15°C.

▼ The Earth's atmosphere lets most of the Sun's heat and light through to the surface. However, the Earth is cooler and radiates energy at longer wavelengths. The atmosphere absorbs much of this radiation and some of this is re-radiated back to Earth. This effect, called the greenhouse effect, keeps the surface of the Earth warmer than it would be if there was no atmosphere

Sun
the Sun warms the Earth's surface
the Earth's surface emits radiation
the atmosphere absorbs and radiates some of the radiation from the Earth's surface
the atmosphere absorbs and reflects some radiation
atmosphere
Earth

▶ Water vapor contributes strongly to the greenhouse effect, even more so than carbon dioxide. On Venus, the high temperatures caused water to evaporate, resulting in more water vapor. This contributed to the greenhouse effect and made the temperatures even hotter. In turn, the hotter temperature caused more water to evaporate, leading to a runaway greenhouse effect. As a result, the atmosphere of Venus is very thick, with surface pressures that are about 90 times greater than the Earth, and temperatures that average about 460°C

How does our atmosphere move?

When the Sun's rays hit the Earth and heat the land, the air above is also heated. This hot air expands and rises. As this air moves upwards, air is drawn in from the side. This effect drives the Earth's winds. The land nearer the equator is hotter because the Sun's rays are more directly overhead and therefore the air gets hotter nearer to the equator. At the poles, however, the Sun's rays hit the Earth at an angle and therefore the land (and the air above it) is cooler. Therefore, the air around the equator is heated and rises, while the air above the poles cools and sinks.

The Earth's rotation breaks this air circulation into atmospheric cells, which provide regular wind patterns around the globe. Other planets also have these patterns—the banded pattern of Jupiter is caused by a similar effect.

ABC An **atmospheric cell** is a region of air circulation where the hot air rises and the cooler air sinks.

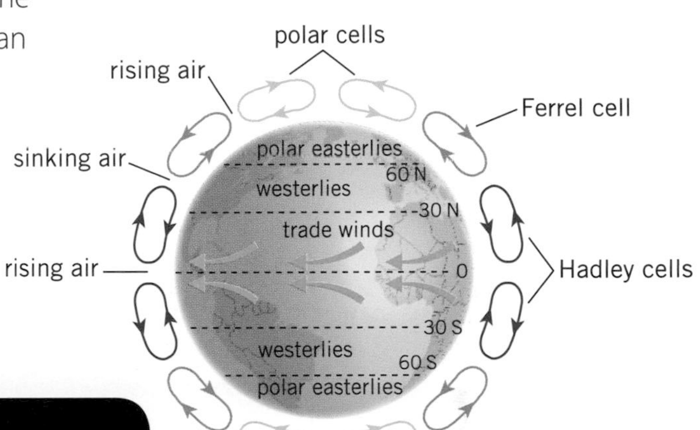

▲ The pattern of rising and sinking air in the atmosphere is driven by energy from the Sun. The Earth's rotation causes these air currents to be broken into cells called Hadley, Ferrel and polar cells. This gives rise to global air circulation and the wind

▲ The bands on Jupiter are caused by the same effect that causes wind circulation on Earth. Because Jupiter is larger than the Earth and rotates faster, the currents are broken into more cells than in the Earth's atmosphere

What is weather?

Although the Earth has some regular wind patterns on a large scale, the wind direction is less reliable on a smaller scale. As the wind moves air around, it creates different weather patterns. Sometimes the wind blows from a cooler part of the globe and brings cold weather, while wind from a different direction can bring warmer weather.

ABC **Transpiration** is the process where water is lost through evaporation from the leaves.

An important consideration in weather is the amount of water vapor in the air. When the Sun's rays hit the ocean, they can cause water to evaporate from the surface. Water is also released through **transpiration** from plants. Higher in the atmosphere, where the temperature is colder, the water vapor can condense and form clouds. Eventually, when there is enough water in a cloud, it will fall as rain or snow.

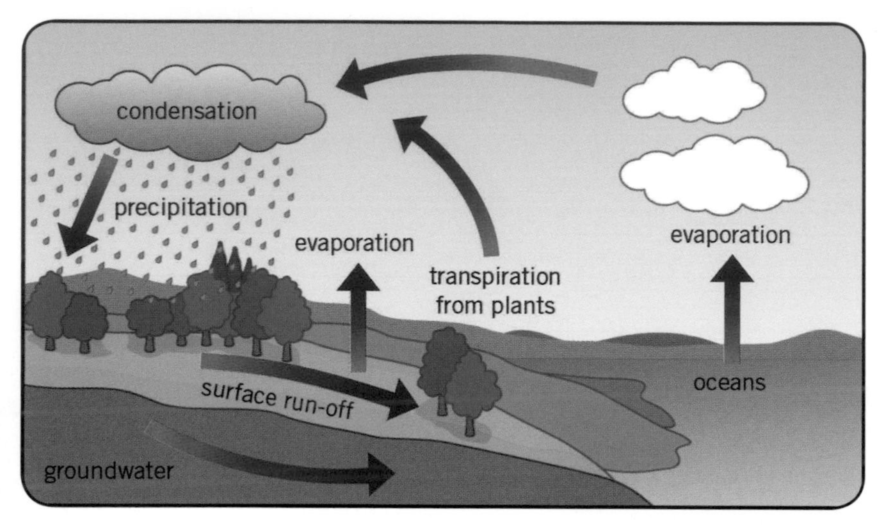

▲ The hydrological cycle, or water cycle, is responsible for providing fresh water to the land

The wind can move clouds over land and therefore much of this rain and snow will fall on land. Some of this water will soak into the ground and be used by plants. Some of the water will run off the surface of the land and form streams and rivers that will eventually return the water to the oceans. These processes are called the water cycle, or hydrological cycle.

Why is the water cycle important?

Without the water cycle, the land would not have water, and life could only exist in the ocean or near the shore. Instead, rain enables plants to grow on land and provide food for animals. Rain is also important as it provides fresh water that does not have salt dissolved in it, unlike seawater. This means that rainwater can be used as drinking water for animals and humans.

Experiment

Investigating evaporation

In this experiment, you will investigate how surface area affects the rate of evaporation. You need a test tube, boiling tube, a small beaker and a larger beaker.

Method

- Using a measuring cylinder, put 10 cm^3 of water in each of the beakers and the test tube and boiling tube.
- Place the test tube, boiling tube and beakers together and leave for 24 hours.
- Use the measuring cylinder to measure the amount of water that is left in each container and record your results.

Questions

1. In which container did the water have the smallest surface area? In which container did it have the largest area?
2. Plot a suitable graph of your data.
3. Why was it important to put the same volume in each container?
4. Why was it important to put the containers near each other when they were left for 24 hours?

Rain is responsible for washing sediment down rivers. In this process, the rivers wash away at their banks in a process called erosion. The action of rivers over thousands of years shapes the land and causes the formation of valleys. The material that is washed from rivers will eventually be deposited at the mouth of the river. As more material

The water cycle causes rain to fall on the land. Some of this water flows off the land in streams and rivers back to the sea

is deposited, the sediment can be compressed to form sedimentary rocks. Sedimentary rocks are discussed further in Chapter 10, The Earth.

Rainwater can also weather rocks directly. As rain falls through the atmosphere, gases such as carbon dioxide dissolve in it. This causes the rainwater to be mildly acidic. Rocks such as limestone or chalk are made from calcium carbonate. This reacts with the acidic rain to form calcium hydrogen carbonate. Calcium hydrogen carbonate is a soluble salt that dissolves in the water and is washed away. This process not only causes the weathering of the rocks but also leaves the calcium salts dissolved in the water.

Tap water that contains a lot of dissolved calcium salts is called hard water. If the water does not have many calcium compounds dissolved in it, then it is soft water. If water that contains calcium hydrogen carbonate is heated and evaporates, calcium carbonate is left behind. This causes limescale, which can block pipes and leave deposits on surfaces where water is able to evaporate.

The water cycle deposits water on the land. As it flows back towards the sea in the form of rivers or glaciers, it carves valleys into the landscape

These limestone rocks are weathered by acidic rain

These stalagmites and stalactites in Mae Usu Cave, Thailand, are caused by acidic rain dissolving some of the rock above. As the water drips into the cave, it evaporates and deposits the compounds that are dissolved in it

1. When hard water evaporates, it can deposit calcium carbonate as limescale. Household cleaners that remove limescale often contain acids.

   a) Why do these cleaning products use acids rather than alkalis?

   b) What hazards are associated with using these cleaning products?

Experiment

Finding the quantity of solutes in water

For this experiment you will need two different samples of water. One could be the water from the tap, the other might be bottled water, rainwater or water collected from a stream.

Method

- Measure the mass of a 250 mL beaker.
- Measure 100 mL of one of your water samples and pour this into the beaker.
- Gently heat the beaker. If using a Bunsen burner, use a medium heat with the beaker placed on a gauze.
- Stop heating the beaker when most of the water has evaporated.
- Leave the beaker to cool and allow the remainder of the water to evaporate.
- Measure the mass of the beaker once all the water has evaporated.
- Repeat the experiment for your other sample of water.

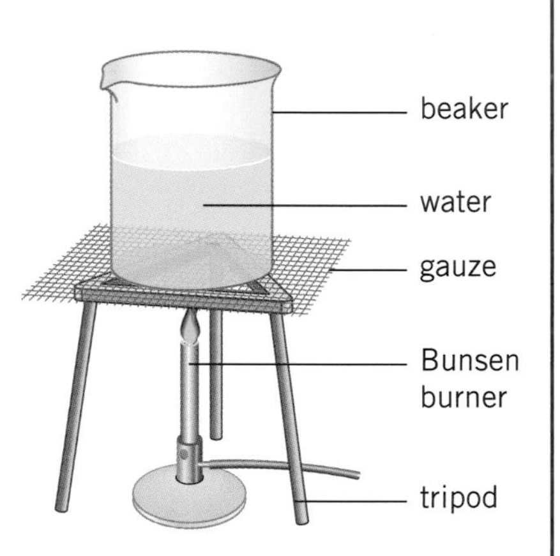

Questions

1. How much extra mass was recorded for each sample?
2. Which sample of water contains the most solutes?

What is climate?

The weather can change every day. There are also seasonal weather patterns that cause the weather to be warmer in summer and cooler in winter. This means that the weather can be different at different times and in different places.

The climate is a measure of the average weather. While different countries will have differing climates, this climate does not change regularly. The climate in a particular location might be affected by altitude, and whether it is close to the equator or closer to the poles. Being near the sea can create milder winter and summer temperatures, while being further inland can give more variation in seasonal temperatures. Other features, such as being near to hills, can change the amount of rainfall.

Features that affect rainfall

The table below shows the average rainfall in mm for some US states. The data is derived from data published by the National Oceanic and Atmospheric Administration, which covers the years 1981 to 2010. The average elevation (height of the land above sea level) of the monitoring stations is also given and if the state borders the ocean, then it is marked as coastal.

State	Elevation (m)	Rainfall (mm)	Coastal	State	Elevation (m)	Rainfall (mm)	Coastal
Alabama	134	1413	Y	Montana	1101	384	
Arkansas	166	1304		New Hampshire	250	1180	Y
California	497	612	Y	New Jersey	62	1207	Y
Delaware	24	1166	Y	North Carolina	284	1263	Y
Florida	16	1386	Y	North Dakota	553	465	
Idaho	1230	442		Oregon	562	934	Y
Illinois	190	1017		Pennsylvania	308	1107	
Kansas	503	799		South Dakota	662	543	
Kentucky	226	1224		Tennessee	262	1360	
Maryland	144	1114	Y	Vermont	292	1136	
Massachusetts	125	1241	Y	Virginia	287	1119	Y
Mississippi	88	1469	Y	Wisconsin	300	845	
Missouri	255	1107		Wyoming	1684	336	

1. Find the average rainfall for the coastal states.
2. Find the average rainfall for the states that do not have a coastline.
3. Do your answers suggest that being nearer the sea affects the amount of rainfall?
4. Plot a graph of rainfall on the *y*-axis vs elevation on the *x*-axis.
5. How does elevation affect rainfall?

How is the climate changing?

The global climate has always been changing slowly. In the past, the average temperature has gone through periods of warmth, as well as ice ages. These changes happen over thousands of years.

Occasionally major events can temporarily affect the Earth's climate. Huge meteorite impacts or large volcanic eruptions can increase the amount of dust at high levels of the atmosphere. These can block some of the sunlight reaching the Earth and cause a shorter period of cooler temperatures.

More recently, studies have shown that human activity is altering the contents of the atmosphere. We are releasing increased amounts of gases such as carbon dioxide, which contribute to the greenhouse effect. As a result, the Earth's climate is getting warmer.

ATL Communication skills

International cooperation

Human activity is releasing greenhouse gases into the atmosphere, which contribute to global warming. All countries must collaborate to reduce the amount of these gases released into the atmosphere and slow the rate of global warming.

The Paris Agreement, which was finalized in 2015, is an agreement that countries can sign up to. The goal of the agreement is to stop the average global temperature from rising by more than 2°C. Countries that sign up must plan to reduce their contribution to global warming and report their progress. By February 2019, 195 countries and member states had signed the agreement.

Measuring carbon dioxide in the atmosphere

Scientists have measured the amount of carbon dioxide in the air for many years. The Mauna Loa Observatory is located in Hawaii at the top of a tall volcano. Scientists use this location to measure the amount of carbon dioxide in the air as its high altitude means that it is above local human sources of carbon dioxide.

The measurements of atmospheric carbon dioxide are shown in the graph on the right. The amount of carbon dioxide varies within each year. The units of the amount of carbon dioxide are parts per million. One part per million means 1 cm^3 of carbon dioxide in one million cm^3 of air (one million cm^3 is one m^3). Hence, 320 parts per million indicates 320 cm^3 per cubic meter of air, which is a concentration of 0.032%.

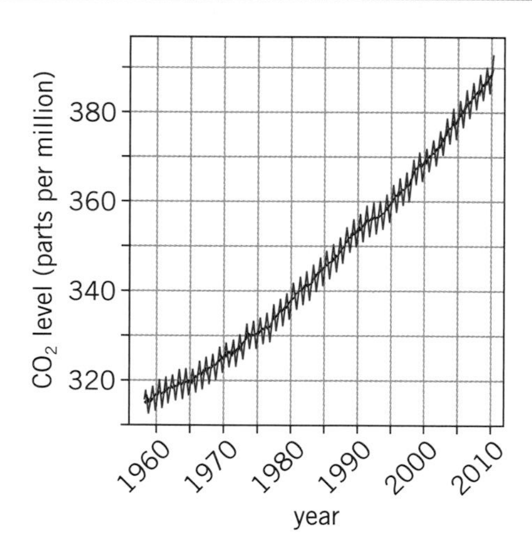

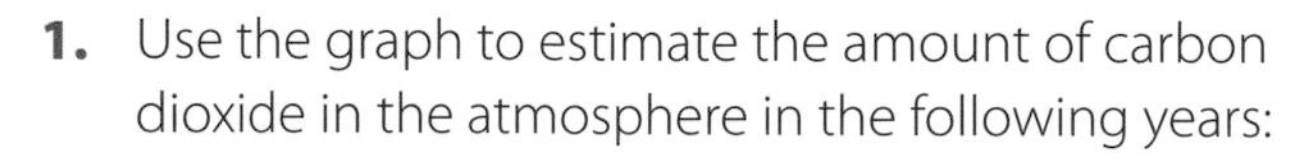

1. Use the graph to estimate the amount of carbon dioxide in the atmosphere in the following years:
   - **a)** 1970
   - **b)** 2000
2. Using your answers to question 1, estimate how much the amount of carbon dioxide increases each year.
3. The monitoring station is located on a volcano, which can emit carbon dioxide. Why is it important to take account of the carbon dioxide emitted by the volcano when measuring atmospheric carbon dioxide?

What are the problems of climate change?

Scientists are keen to predict what might happen if the average temperatures on Earth rise. There are large amounts of ice in Greenland and Antarctica that could melt if temperatures rise too much. The resulting water would flow off the land and into the oceans, causing the sea levels to rise. Sea levels are currently rising by about 3 mm per year, but this might increase such that sea levels rise by 2 m or more over 100 years. The increase in sea levels will depend on how much global temperatures rise.

If sea levels rise too much, then people living near the coast will be affected. Many cities are located near to the sea. Rising sea levels will require people living in these cities to relocate, and significant flood defenses may need to be built.

1. If sea levels continue to rise at 3 mm per year, how much would the levels rise over 100 years?
2. If sea levels rise by 2 m in 100 years, how many millimeters would sea levels rise on average per year?

A changing climate will result in many regions of the Earth becoming hotter. Some areas, particularly those inland, will become drier and have more extreme seasonal temperature changes. As a result, these areas would be hotter in summer but cooler in winter. Coastal areas might experience warmer temperatures but with increased rainfall and milder winters.

As a changing climate affects weather patterns, land that was once used for farming might no longer be suitable for this purpose. Farmers will have to adapt by growing different crops in different parts of the world.

▼ If the sea level rises too much, coastal cities such as New York will have a problem

Summative assessment

Statement of inquiry:

The atmosphere around us creates the conditions necessary for life.

This assessment is based on the movement of the atmosphere and how wind can be used to generate power.

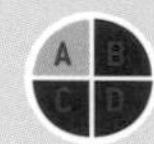

Properties of the atmosphere

1. The wind stores energy in which form?
   - **A.** Gravitational potential
   - **B.** Heat
   - **C.** Kinetic
   - **D.** Sound
2. What is the original source of this energy?
   - **A.** Heat energy from inside the Earth
   - **B.** The orbit of the Earth around the Sun
   - **C.** The Sun
   - **D.** The Moon
3. What is the most common element in the wind?
   - **A.** Argon
   - **B.** Nitrogen
   - **C.** Oxygen
   - **D.** Water
4. Winds in the stratosphere can have a higher speed than those nearer the surface of the Earth. Which of the following reasons explains why the energy carried by these winds is lower?
   - **A.** The winds are at a higher height.
   - **B.** The wind is at a higher temperature.
   - **C.** The wind is at a lower temperature.
   - **D.** The air has a lower density.
5. Most weather occurs in the troposphere. Which change in the atmosphere occurs as you go to a higher altitude in the troposphere?
   - **A.** The density increases
   - **B.** The temperature increases
   - **C.** The amount of nitrogen increases
   - **D.** The pressure decreases

Investigating a model wind turbine

An engineer is investigating how the power output of a wind turbine depends on the wind speed. She decides to make a model of a wind turbine for this investigation. She uses a fan to generate a flow of air, which hits the blades of a model wind turbine. The fan has a variable power supply so that the speed of the air can be changed. The model wind turbine turns a small electric generator and the output power from the generator is recorded. A schematic diagram of the apparatus is shown below.

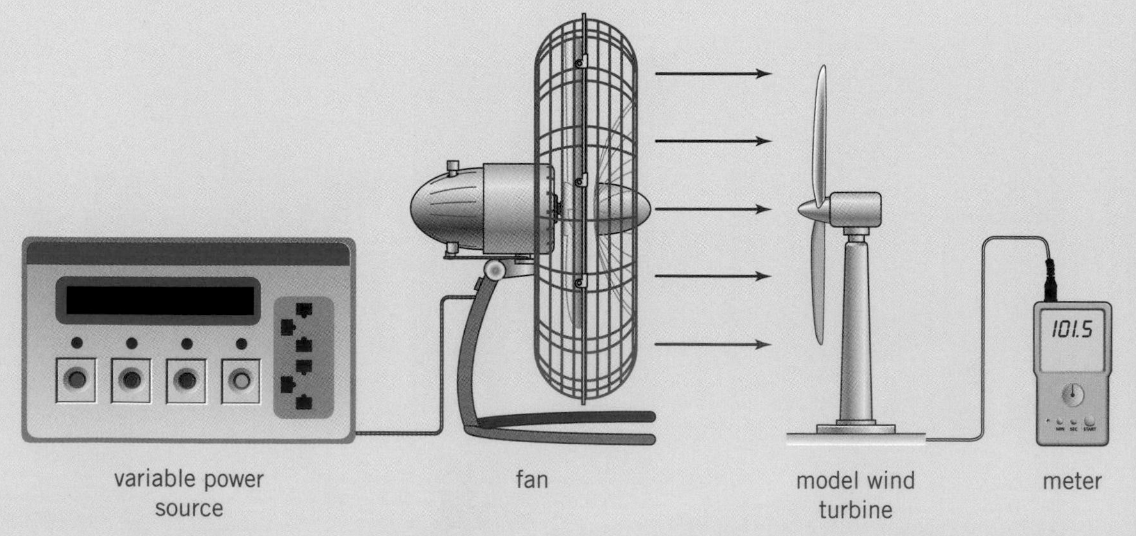

6. Describe the advantages and disadvantages of using a model wind turbine rather than using an industrial wind turbine for this investigation. [4]
7. The engineer has a device called an anemometer, which measures the air speed generated by the fan. Describe a method that the engineer could follow in order to measure how the power output of the fan changes with the speed of the air. [4]
8. As part of further investigations, the engineer wants to investigate other factors which affect the power output of a wind turbine. Suggest a different factor that might affect the power output. [2]

Factors affecting the output of wind turbines

The table on the right shows the radius of the arms of different makes of wind turbines and the power output of the turbines.

Radius of wind turbine arms (m)	Power output (MW)
70	1.5
80	1.8
90	3.0
120	4.1
190	11.0

9. Plot a graph of the data in the table. [4]
10. Add a line of best fit to your graph. [1]
11. The wind turbine with a radius of 190 m is yet to become commercially available. The largest available wind turbines have a radius of about 150 m. Use your graph to estimate the power output of this type of turbine. [2]
12. The data in the table suggest that larger turbines generate more power. Describe the disadvantages of building large wind turbines. [3]

Wind turbines

13. Wind turbines are an important source of renewable energy.
    a) Explain what is meant by renewable energy. [2]
    b) Describe the problems associated with non-renewable sources of energy. [3]
    c) Describe the disadvantages of the use of wind turbines to solve the problems of non-renewable energy. [2]
    d) Describe the energy transfers that occur when a wind turbine is used to supply power. [3]

◀ Wind turbines are an important source of renewable energy

12 The universe

Key concept: Relationships

Related concepts: Evidence, Movement

Global context: Orientation in space and time

The Antikythera mechanism was recovered from an ancient Greek shipwreck in 1901. The shipwreck is believed to have occurred in 70–60 BC and therefore the mechanism must be older than this. It appears that the mechanism was a mechanical device that could be used to calculate the position of the Moon and the Sun and even calculate the occurrence of eclipses. Why did ancient civilizations want to measure the position of the Sun and Moon?

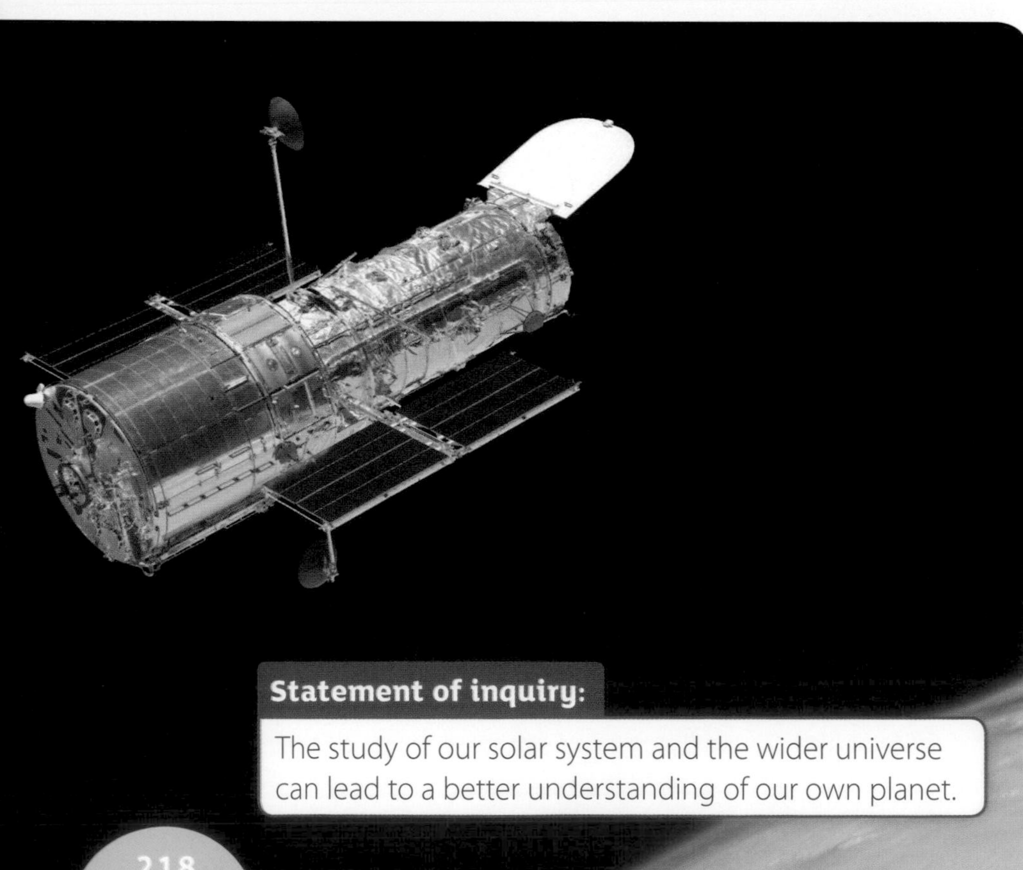

Telescopes have been fundamental to our discovery and exploration of space because they enable us to see things that are too far away or too dim for our eyes to see. The Hubble Space Telescope was launched in 1990 and has an advantage over telescopes that are on the ground because the Earth's atmosphere does not distort the image. The Hubble Space Telescope has provided us with some of the most important images of objects in space that we have. Why is seeing things so important as a way of understanding what is happening?

Statement of inquiry:

The study of our solar system and the wider universe can lead to a better understanding of our own planet.

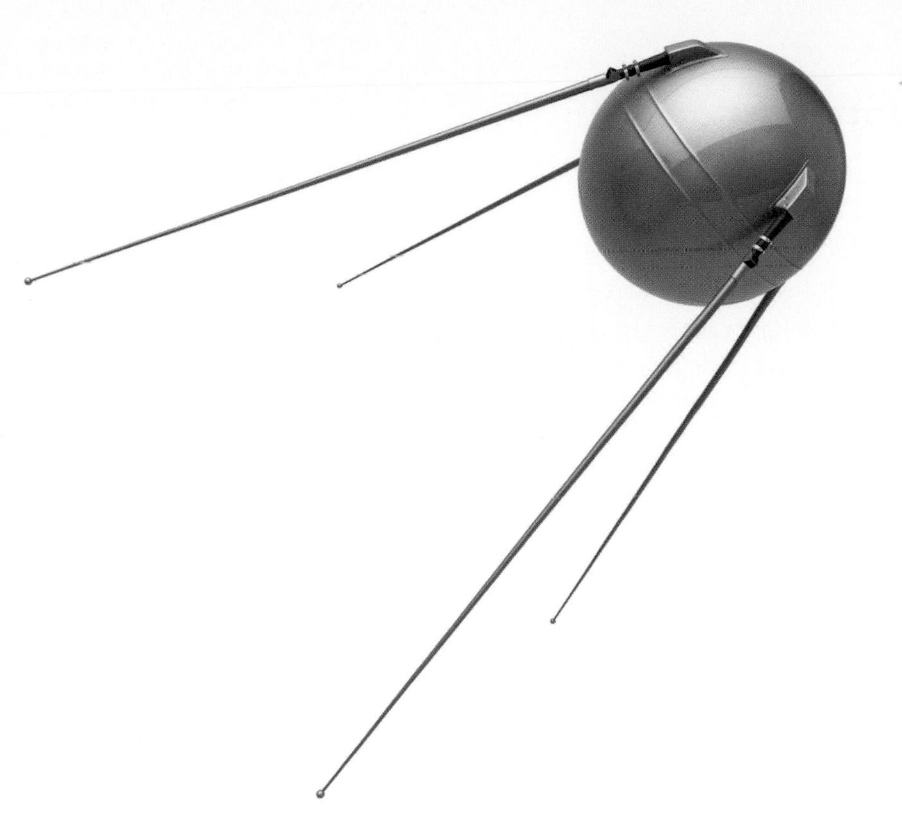

Sputnik was a Russian satellite, which was the first artificial object in space. It orbited the Earth for three months in 1957, before falling back into the Earth's atmosphere. The success of Sputnik triggered the Space Race. The United States and Soviet Union rivaled each other to develop the technology to send satellites into space, and to the Moon, Venus and Mars. It eventually led to the first human landing on the Moon in 1969. How do we use satellites today?

Exploring and investigating space requires investment in technology. While this technology can be expensive, it often finds other uses. Technology that was developed for space exploration has given us new materials that can be used for insulation, more efficient ways of generating renewable energy, better medical equipment and improved computing. NASA research into how to provide nutrition for astronauts on long journeys led to the development of important additives to infant formula. What other examples are there of spin-off technology that were developed for space exploration?

Key concept: Relationships

Related concepts: Evidence, Movement

Global context: Orientation in space and time

Statement of inquiry:

The study of our solar system and the wider universe can lead to a better understanding of our own planet.

Introduction

From the very earliest times in history, humans have looked at the sky and tried to interpret what they were seeing. For many people, the explanations and stories that explained the patterns in the stars and the motion of the planets were part of the same story that explained how the Earth came to be here. People also believed in horoscopes, which used the positions of the planets to forecast the future.

While scientists no longer believe that the alignment of planets at the time of our birth affects our futures, or that the Sun is chasing the Moon across the sky, we are still interested in studying the planets and stars because it helps us to explain how the Earth might have formed. In this chapter, we will see where the Earth sits in the solar system and within the universe. The global context of this chapter is orientation in space and time.

One of the challenges of astronomy is that everything is very far away, and things tend to change very slowly over millions of years. This makes it difficult to collect evidence about how the universe works. In this chapter, we will see how we have arrived at our ideas of how the universe began and what might happen in the future. The related concepts of the chapter are evidence and movement.

▶ The Sun appears to have been of great importance to ancient cultures and civilizations. Many monuments appear to have been built so that they align with the Sun on a particular day such as the summer solstice—the day on which the Sun reaches its highest point in the sky. Egyptian pyramids and Stonehenge are examples of this. Machu Picchu is a citadel built by the Inca people and later abandoned. In it, the Torreon, or Temple of the Sun, is built so that it aligns with the rising Sun on the summer solstice. While the purpose of many of these buildings is unknown, it is thought that they were used as a sort of calendar

What is in the solar system?

The Sun is by far the biggest object in the solar system. It is more than 1,000 times heavier than Jupiter (the second biggest object in the solar system) and more than 330,000 times heavier than the Earth. As a result, the Sun dominates other objects with its gravitational field. These other objects will experience a force towards the Sun and will accelerate towards the Sun. Instead of falling into the Sun, the sideways motion of these objects keeps them orbiting around the Sun.

The surface of the Sun has a temperature of about 6,000°C. In the center of the Sun, however, the temperature is about 15 million degrees Celsius. This huge temperature and the enormous pressure in the center of the Sun enable it to convert hydrogen into helium. This is the way that the Sun generates its power.

By converting hydrogen into helium, it is converting mass into energy according to Einstein's famous equation, $E = mc^2$. Because of this, the Sun loses about 4 million tonnes of mass every second. Although this is a large change in mass, the Sun is so vast that over its lifetime of about 10 billion years it will only lose about $\frac{1}{15}$th of a percent of its mass.

Viewing sunspots

The Sun has a strong magnetic field. Sometimes this field can cause cooler patches in the surface of the Sun, which appear darker. These are called sunspots. It is possible to view sunspots, but you must never try this by looking directly at the Sun. Doing so could damage your eyesight. Instead, you can create an image of the Sun on a piece of paper.

This activity needs to be done on a sunny day. To create the image of the Sun, you will need a piece of paper to project the image onto. You can use a small telescope or one side of a pair of binoculars. Do not look through the telescope or binoculars, but point them towards the sun and put the piece of paper a small distance from the eyepiece. Adjust the angle of the binoculars so that Sun's light shines through onto the paper. Then move the paper towards or away from the eyepiece to focus the image. You should be able to create an image of the Sun and if there are any large sunspots you may be able to see them.

▲ Sunspots on the surface of the Sun

What makes a planet?

After the Sun, the next biggest objects in the solar system are the **planets**. There are eight objects that are recognized as planets. In order of distance from the Sun, they are Mercury, Venus, Earth, Mars, Jupiter, Saturn, Uranus and Neptune. These planets are different from other objects in the solar system in the following ways:

- They are spherical (unlike asteroids, which have irregular shapes).
- They orbit the Sun (as opposed to moons, which orbit planets).
- They are the largest object in that orbit (unlike an asteroid, which might be just one of many).

These form the definition of a planet. There are two other features of the planets in our solar system:

- Their orbits are circular (as opposed to comets, which have very elliptical orbits).
- They all orbit in the same plane.

ABC

A **planet** is one of eight major objects in the solar system that orbit around the Sun.

ABC An **inner planet** is one of the smaller planets that are closer to the Sun than the asteroid belt. They are Mercury, Venus, Earth and Mars.

A **terrestrial planet** is one that is rocky. The terrestrial planets are Mercury, Venus, Earth and Mars.

The planets also fall into two distinct groups. The **inner planets**, Mercury, Venus, Earth and Mars, are all close to the Sun and are made of rock with an iron core. Earth is the largest of these inner planets. They are sometimes called **terrestrial planets** because they are all similar to the Earth.

▲ The two brightest stars in this photo are not stars at all, but planets. Venus is above and to the right of Jupiter. Unlike the stars in the picture, which may be hundreds of light years away, Venus and Jupiter are much smaller, much closer and do not emit their own light. Instead, we see them by the light that they reflect from the Sun

ABC An **outer planet** is one that is beyond the asteroid belt. The outer planets are Jupiter, Saturn, Uranus and Neptune.

A **moon** is an object that orbits around a planet.

The **outer planets** (Jupiter, Saturn, Uranus and Neptune) are very different. They are much further away from the Sun and bigger than the inner planets. The lightest of the outer planets is still almost 15 times heavier than the Earth and four times the diameter.

The only inner planets to have **moons** are Earth, which has one moon, and Mars, which has two small moons. The outer planets, on the other hand, have many moons each. Jupiter currently has 79 known moons, although not all of them have names. The largest of these moons, Ganymede, is larger than Mercury.

▲ When drawn to scale, it is clear that the gas giant planets are much bigger than the terrestrial planets, and the Sun is vastly bigger than these. This picture cannot show the relative distances between the planets on the same scale. If it did, the Earth would be about 14 meters from the picture of the Sun and Neptune would be over 400 m away

Planetary orbits

The data in the table below shows the distance of the first five planets from the Sun and the length of a year on that planet (the amount of time that it takes for the planet to orbit the Sun). The distance from the Sun is given in astronomical units (AU). 1 AU is the distance from the Earth to the Sun.

Planet	Distance from the Sun (AU)	Length of a year (Earth years)
Mercury	0.39	0.24
Venus	0.72	0.62
Earth	1.00	1.00
Mars	1.52	1.88
Jupiter	5.20	11.86

1. Plot a graph of the data on a copy of the axes below.

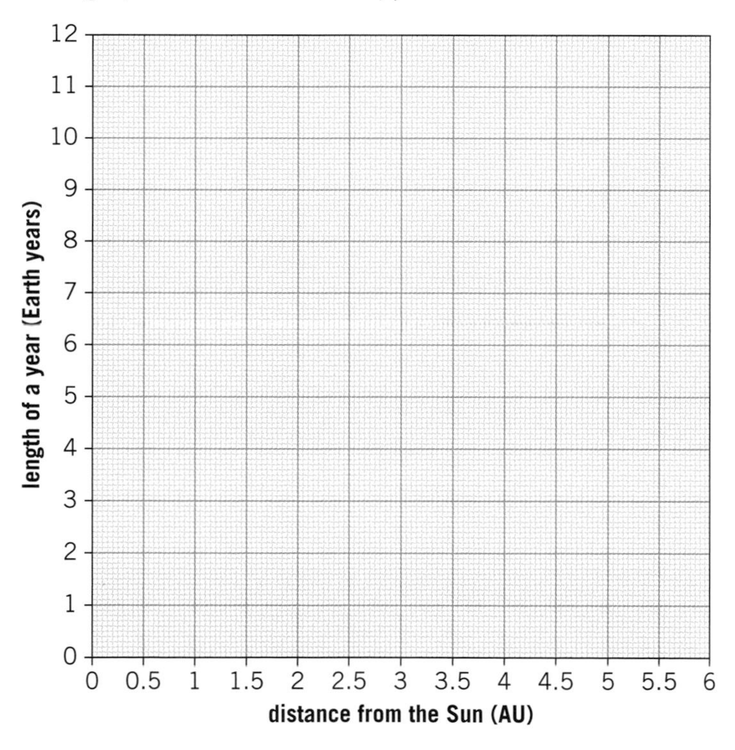

2. Describe the trend of this graph.
3. Ceres is the largest asteroid in the asteroid belt. It takes 4.6 Earth years to orbit the Sun. Use your graph to find the distance of Ceres from the Sun.
4. Vesta is another asteroid that is 0.4 AU closer to the Sun than Ceres. How much less time does it take to orbit the Sun than Ceres?

ABC

An **asteroid** is a rocky object that is too small to be classed as a planet. Many asteroids orbit the Sun in a region between the orbits of Mars and Jupiter called the **asteroid belt**.

ABC A **dwarf planet** is an object that orbits the Sun and is large enough to become spherical in shape. However, it is not large enough to be classed as one of the major planets.

The **Kuiper belt** is a region of space beyond Neptune's orbit that contains a large number of asteroids.

What objects in the solar system are smaller than a planet?

Beyond the orbit of Mars is the **asteroid belt**. This is a collection of millions of small rocky objects. There are estimated to be between one and two million **asteroids** that are more than a kilometer in size and there will be many more millions of asteroids that are smaller. The largest asteroids can be hundreds of kilometers across. The largest asteroid, Ceres, is big enough to be spherical in shape and was even once thought to be a planet.

Although it is the largest object in its orbit, Ceres is no longer thought to be a major planet. This is because it shares its orbit with many other asteroids, some of which are also quite large. It does not dominate the region of space in the same way as the major planets do. Instead, it is defined as a dwarf planet. A **dwarf planet** is one that is large enough to become spherical in shape, but not large enough to dominate and clear its orbit.

▲ NASA's *Dawn* mission sent a probe to image the two largest objects in the asteroid belt: Ceres and Vesta. The probe arrived at Ceres in 2015 and this picture confirms that it is spherical in shape. Therefore it is classed as a dwarf planet

Another object that has been classified as a dwarf planet is Pluto. From its discovery in 1930, it was thought of as the ninth planet; however, the discovery of other objects that had equal claim to be planets caused Pluto, and some of these other objects, to be reclassified as dwarf planets in 2006.

We now know that Pluto is just one object among many others in a region of space called the **Kuiper belt**. This is similar to the asteroid belt but it is situated beyond the orbit of Neptune. Astronomers have found about a thousand objects in the Kuiper belt, some of which are larger than 1,000 km across. There are likely to be millions more objects that are smaller in size.

How have we explored the solar system?

▲ Exploring the solar system is an important way of investigating it. Even with the best telescopes, we are limited in what we can see by the large distances of objects in the solar system. The picture on the left shows one of the best images that we had of Pluto, taken by the Hubble Space Telescope (which is in orbit around the Earth). The image is good enough to see that Pluto and its moon Charon have two smaller moons (Nix and Hydra), which are the two smaller spots to the right of the two larger spots. When the *New Horizons* probe reached Pluto in 2015, it was able to take pictures such as the one on the right, which shows Pluto and Charon in much more detail

When studying the solar system, astronomers are limited in that they are studying objects that are a long way away. The development of the **telescope** in the 17th century enabled astronomers to see these distant objects more clearly.

Without telescopes, the only planets other than Earth that were known were Mercury, Venus, Mars, Jupiter and Saturn. It was only possible to discover more distant objects with the use of telescopes. Uranus was discovered in 1781 and Neptune in 1846.

For some planets far away from Earth, it was necessary to send **probes** in order to get good pictures and take measurements, rather than rely on telescopes. *Pioneer 10* was launched in 1972 and investigated Jupiter. This was closely followed by *Pioneer 11* in 1973, which studied Jupiter and Saturn.

In the late 1970s, there was an opportunity to send probes to all of the gas giant planets. This was due to these planets aligning in a way that would assist the journey of space probes. This alignment occurs once every 175 years and the Voyager missions were launched in 1977. *Voyager 1* visited Jupiter and Saturn, while *Voyager 2* visited Jupiter, Saturn, Uranus and Neptune. More than 40 years after they were launched, the Voyager probes are still operational and are the most distant human-made objects from Earth.

While flying a probe past planets gives good pictures and useful information about planets, it is still difficult to examine the surface directly, or to investigate what is just underneath the surface of planets. To do this, it is necessary to land on the planet or moon. The first human-made object to land on the Moon was *Luna 2*, which crashed into the Moon in 1959. It took measurements on its way to the Moon's surface, but the crash landing prevented it from sending data once it was on the Moon.

The first time a probe had a soft landing on the moon was in 1966, when *Luna 9* landed and was able to send back pictures of the Moon's surface. In 1969, NASA's Apollo 11 mission successfully landed two men on the Moon and brought them back to Earth. The Apollo missions successfully achieved five further Moon landings and landed 12 astronauts on the Moon in total. The missions brought back many samples of moon rock.

Humans have never traveled further than this. However, space probes have successfully landed on more distant objects. Four **rovers** have successfully landed on Mars but were not designed to return samples. Other missions have successfully returned samples from a **comet** (Wild 2) and an asteroid (25143 Itokawa).

ABC

A **telescope** is an optical instrument that makes distant objects appear nearer and brighter.

A **probe** is a satellite, usually with measuring equipment, that is sent into space.

A **rover** is a vehicle that is designed to travel around on another planet or moon, usually to take measurements or pictures.

A **comet** is an icy rock that travels in an elongated orbit that brings it close to the Sun for a short period of time. When it is close to the Sun, the ice evaporates and forms a tail of gas and dust.

A self-portrait of the *Curiosity* rover on Mars

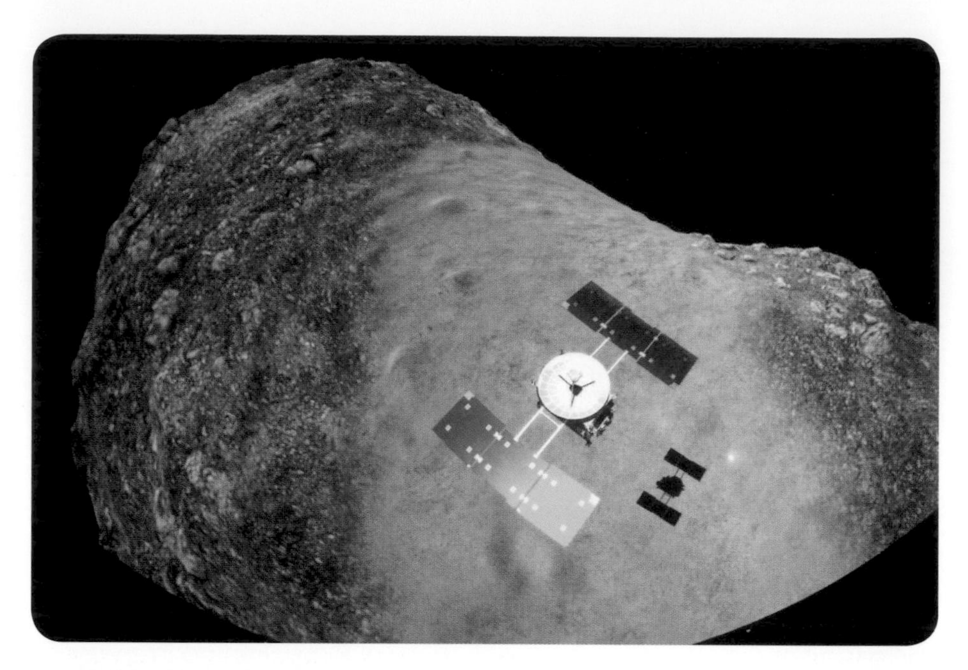

▲ The Japanese *Hayabusa* mission landed on the asteroid Itokawa in 2005 and returned samples of rock to Earth in 2010. These asteroid samples help scientists to determine the nature of the early solar system and how it formed. The *Hayabusa2* mission is expected to return samples from another asteroid in December 2020

What lies beyond the solar system?

The *Voyager 1* space probe is now the furthest human-made object from Earth. It is over 20 billion km from the Earth, and it takes light more than 20 hours to get between the Earth and the probe. It has been traveling since 1977 and its speed is 61,000 km per hour. However, even at this speed it would take tens of thousands of years to reach the nearest stars.

To measure the vast distances in space, astronomers often use the unit of **light years**. A light year is the distance that light travels in a year. The speed of light is 300 million m/s and so 1 light year is 9.5 million million km (9.5×10^{15} m).

ABC A **light year** is a unit of distance used to measure the large distances in space. It is the distance that light travels in one year.

An **exoplanet** is a planet that is found in another solar system orbiting a star other than our Sun.

The second-nearest star to us is Proxima Centauri, which is 4.2 light years from Earth. Proxima Centauri is a small star with a radius that is only 15% that of the Sun and a mass that is eight times lower than the Sun's mass. As a result, it emits about 20,000 times less visible light than the Sun.

In 2016, astronomers discovered a planet that orbits around Proxima Centauri. This planet orbits about 20 times closer to Proxima Centauri than the Earth orbits the Sun. Although this planet is closer to its star than Earth is, Proxima Centauri is also much cooler than the Sun. This makes it possible that the planet has liquid water on its surface.

Astronomers first confirmed the existence of planets orbiting other stars that were like our Sun in 1995. Since then, about 4,000 **exoplanets** have been confirmed. Many astronomers believe that most stars will have planets orbiting around them.

1. The speed of light is 300 million m/s. Calculate the distance that light travels in
   a) 1 hour
   b) 1 minute
   c) 1 day.
2. Most of the planets that have been discovered have been gas giants like Jupiter, rather than terrestrial planets like the Earth. Why do you think that it is easier to detect gas giant planets?

Extra-terrestrial life

Astronomers are interested in planets that are like Earth because there is a possibility that they might have life on them. If a planet is too close or too far away from its star, then the temperature might be too hot or too cold for water to exist as a liquid. Planets that orbit at a suitable distance from their star for liquid water to exist are said to be in the habitable zone, or the "Goldilocks zone".

1. How might the Goldilocks zone be different for a large hot star compared to a small cool star?
2. If life were to exist on a planet that had a higher gravitational field strength than Earth, how might it differ from life on Earth?

ATL Communication skills

Making contact

Many people have thought about how they might communicate with extra-terrestrial life. Such a life form would not understand our language, our number system or even the units in which we measure things. Trying to communicate which planet we come from, or how big we are, would be difficult.

When the Voyager probes were launched in 1977, each one carried two golden records. They contain spoken greetings in many different languages and sounds, along with music and pictures.

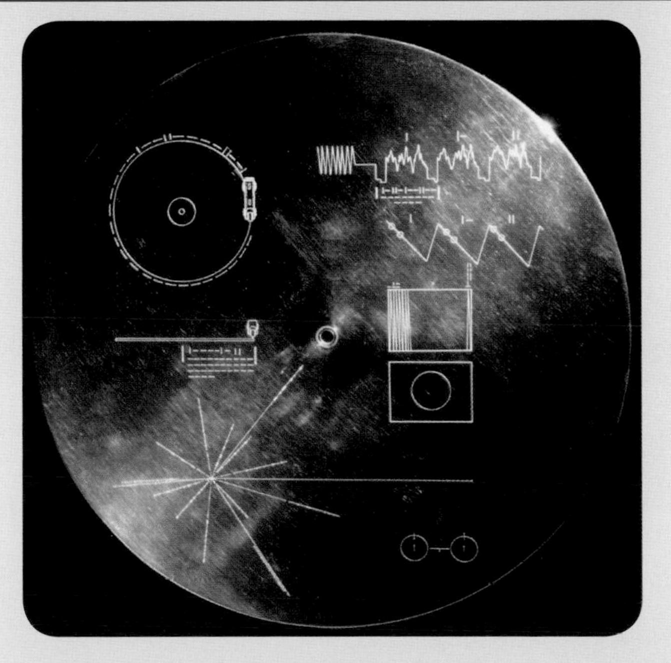

▲ The cover of the Voyager Golden Record

A **galaxy** is a collection of billions of stars held together by the force of gravity.

The **Milky Way** is the galaxy in which our Sun can be found.

What does our galaxy look like?

Our star is just one of hundreds of billions of stars in the **Milky Way galaxy**. We believe that the Milky Way is a spiral shape. However, because we are inside the galaxy, we cannot see its shape. The galaxy is about 100,000 light years across and 1,000 light years thick. This vast size means that the likelihood of us ever being able to send a probe outside the galaxy is very low.

▶ Although we cannot directly see the shape of our own galaxy, we can see many other galaxies that have a spiral shape. The Whirlpool Galaxy is about 23 million light years away

Astronomers have found that at the center of almost all galaxies, even the Milky Way, is a **supermassive black hole**. A black hole is an object that is so dense, and its gravitational field so strong, that nothing—not even light—can escape if it gets too close. The possibility of black holes has been known about for many centuries but detecting them is difficult because very little light can escape from them.

▲ The first ever image of a black hole was captured by the Event Horizon Telescope in April 2019. It shows the supermassive black hole at the center of the Messier 87 galaxy

Some black holes form when a large star explodes in a supernova, but the black holes in the center of galaxies are millions of times heavier. This is why they are referred to as supermassive. The black hole at the center of the Milky Way is called Sagittarius-A* and is thought to have a mass that is about 4 million times larger than the mass of the Sun.

What lies beyond our galaxy?

The Milky Way is not the only galaxy. It is part of the **Local Group** of galaxies, which consist of about 50 dwarf galaxies and three large spiral galaxies: the Milky Way, Andromeda and Triangulum. These galaxies are 2.5–3 million light years away.

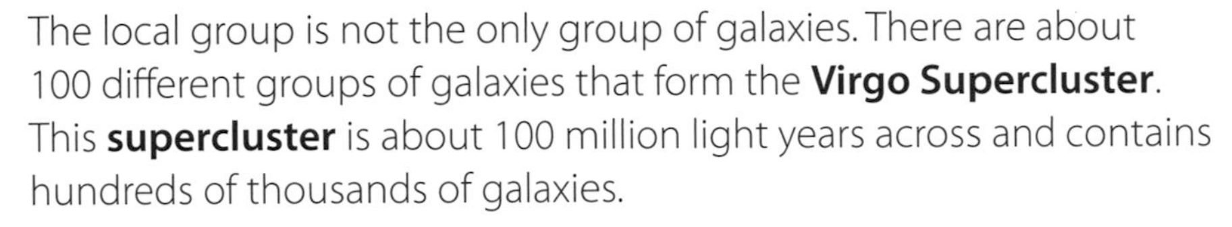

The local group is not the only group of galaxies. There are about 100 different groups of galaxies that form the **Virgo Supercluster**. This **supercluster** is about 100 million light years across and contains hundreds of thousands of galaxies.

What is at the edge of the universe?

A supercluster is one of the largest structures in the universe. These superclusters are approximately 100 million light years across, and the nearest superclusters are about 500 million light years away. Astronomers believe that there are about ten million superclusters in the observable universe.

The furthest superclusters of galaxies are about ten billion light years away. This means that the light from these galaxies has been traveling towards us for most of the history of the universe. Much further than this, objects are so distant that the light has not had enough time to reach us. As a result, we cannot see these objects yet and they are outside of the observable universe.

Astronomers believe that the universe extends beyond the edge of the observable universe for vast distances. Some studies suggest that it extends at least hundreds of times further than the observable universe, others suggest that it could be infinite.

ABC

A **black hole** is a region of space containing an object of infinite density. The gravitational field of a black hole is so large that not even light can escape from it.

A **supermassive** black hole is one that has a mass that is millions of times greater than the Sun. Supermassive black holes are found in the center of galaxies.

A **group** of galaxies is a collection of galaxies that are held together by their gravity.

The **Local Group** of galaxies is the galactic group that contains the Milky Way.

This picture, called the Hubble eXtreme Deep Field, was released in 2012. It is a composite of images taken by the Hubble Space Telescope over a ten-year period, and shows some of the most distant objects ever seen. The dimmest galaxies in the picture are billions of times dimmer than what our eyes can detect. The light from the furthest galaxies has taken about 13.2 billion years to reach the Earth, which is most of the age of the universe. Images like this demonstrate just how large the universe is

ABC A **supercluster** is a collection of hundreds of thousands of galaxies.

The **Virgo Supercluster** is the supercluster of galaxies containing the Milky Way.

How did the universe start?

In 1929, astronomer Edwin Hubble found that all distant galaxies are moving away from us in a way that suggests that they all started from the same place. This observation led to the idea that the universe started in at a single point and exploded outwards. This theory is called the Big Bang. By measuring the current motion of galaxies and how far they are, astronomers have tracked their motion backwards and conclude that the Big Bang occurred about 13.8 billion years ago.

At the time of the Big Bang, the entire universe would have been incredibly small, hot and dense. We do not yet understand the laws of physics that are needed to describe such a universe. In the first few seconds after the Big Bang, the universe was too hot for protons and neutrons to form. As the universe cooled, protons and neutrons formed and later formed atoms. The first stars and galaxies, including the Milky Way, formed about 300 million years after the Big Bang. Our Sun formed about 9 billion years after the Big Bang.

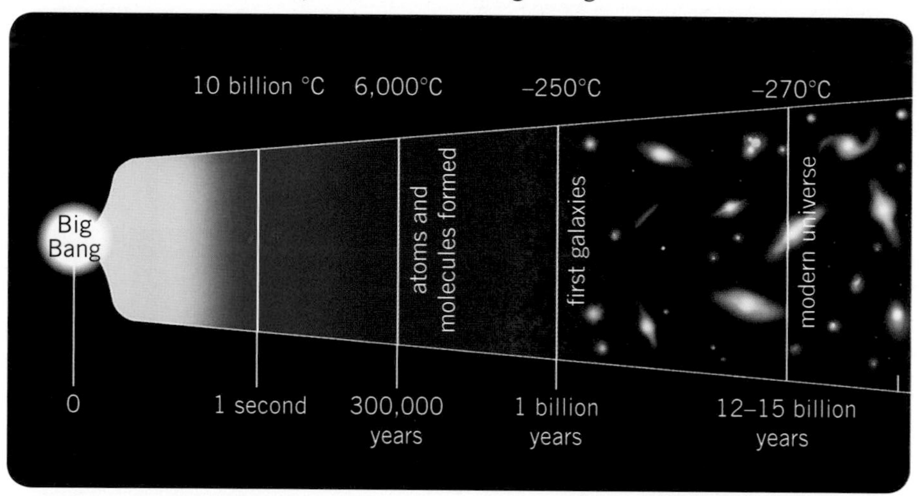

▲ This diagram shows the history of the universe since the Big Bang

Creating a timeline of the universe

Use some of the information in the table to create a timeline of the universe.

Time (billions of years)	Event
0	Big Bang
0.0004	atoms form
1	Milky Way galaxy forms
9	Sun and Earth form
10	first life on Earth
13.5	first dinosaurs
13.7	dinosaurs become extinct
13.769	first humans
13.77	present day

Summative assessment

Statement of inquiry:

The study of our solar system and the wider universe can lead to a better understanding of our own planet.

This assessment is based on the search for exoplanets and the potential for finding extra-terrestrial life.

Objects in the universe

1. An exoplanet is a planet that orbits
   - **A.** another planet
   - **B.** another star
   - **C.** another galaxy
   - **D.** another moon.

2. The temperature of an exoplanet is an important factor that might determine whether life can exist on it. Which of these factors does not affect the temperature of the planet?
   - **A.** The brightness of the star
   - **B.** The size of the planet
   - **C.** The distance of the planet from the star
   - **D.** The exoplanet's atmosphere

3. In 1974, the Arecibo observatory broadcast a message in the direction of a cluster of stars 25,000 light years away. The signal travels at the speed of light. If an extra-terrestrial life form in this star cluster detects the message and sent a reply, in which year would we detect the response?
   - **A.** 6974
   - **B.** 14474
   - **C.** 26974
   - **D.** 51974

4. The Voyager space probes carry a Golden Record, which has pictures and sounds from Earth on it. The record is intended for any life-form that finds it in the future. In 2010, the *Voyager 1* probe was 17 billion km from the Sun and in 2018 it was 21 billion km from the Sun. How fast is the *Voyager 1* probe traveling?
   - **A.** 0.5 km/s
   - **B.** 4 km/s
   - **C.** 13.6 km/s
   - **D.** 15.9 km/s

5. Which of the following is not a reason why we cannot detect exoplanets directly?
   - **A.** Exoplanets are too small.
   - **B.** Exoplanets do not emit their own light.
   - **C.** Exoplanets are too far away.
   - **D.** Exoplanets are too cold.

6. Which is the correct order from smallest to largest?
   - **A.** Exoplanet, Galaxy, Star, Asteroid
   - **B.** Asteroid, Exoplanet, Star, Galaxy
   - **C.** Star, Galaxy, Exoplanet, Asteroid
   - **D.** Galaxy, Exoplanet, Asteroid, Star

Investigating the Goldilocks zone

A pupil wishes to investigate the Goldilocks zone and how the temperature of a planet varies with the distance from a star. She suggests the following method for the experiment.

- Light a bonfire.
- Place a football 1 m away from the bonfire.
- Using a remote thermometer, measure the temperature of the surface of the football that is facing the bonfire.
- Place a second football at a distance of 2 m from the bonfire and measure the temperature of its surface.
- Repeat this method with footballs placed every meter away from the bonfire up to 10 m from the bonfire.

7. From this method, suggest:

a) the independent variable [1]

b) the dependent variable. [1]

8. Suggest one piece of measuring equipment that the pupil will need, other than a thermometer. [1]

9. Describe a control variable that the pupil has successfully controlled. [2]

10. Describe a variable that the pupil should control but has not. [2]

11. Suggest an improvement to make the experiment safer. [2]

Detecting exoplanets

Exoplanets can be detected if they pass between us and the star that they orbit. As they pass, they block some of the light from the star. To detect the exoplanet, astronomers monitor the light from the star and look for regular dips.

Graphs 1 and 2 at the top of page 233 each show dips in the light intensity of stars like our Sun. In both cases, this is evidence for the existence of an exoplanet.

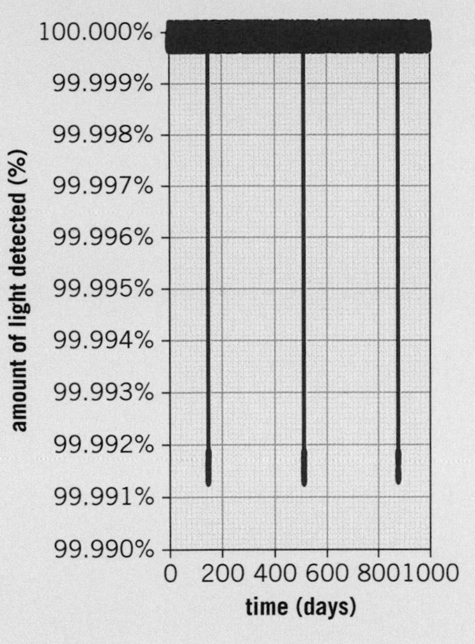

Graph 1

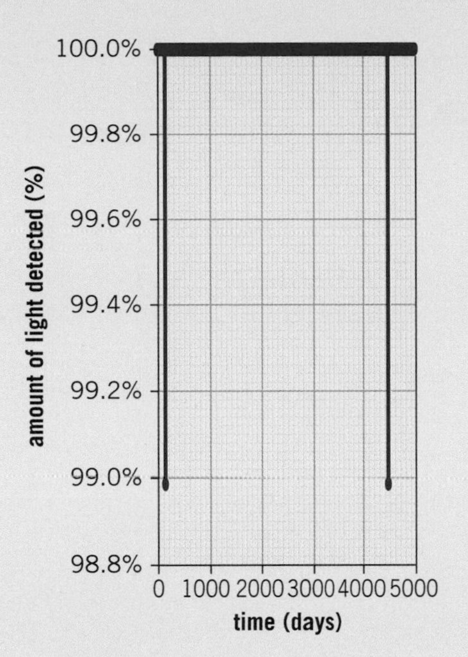

Graph 2

12. State two differences between the graphs. [2]

The amount by which the light dips below 100% can give information about the size of the exoplanet, as a larger planet blocks more of the light. The graphs below show how the amount by which the light dips relates to the size of the planet.

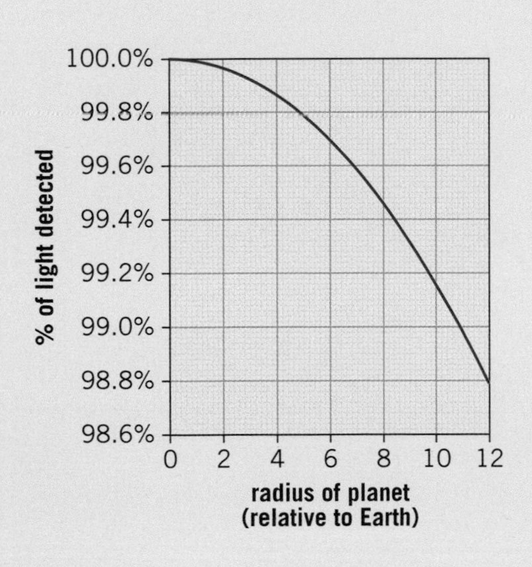

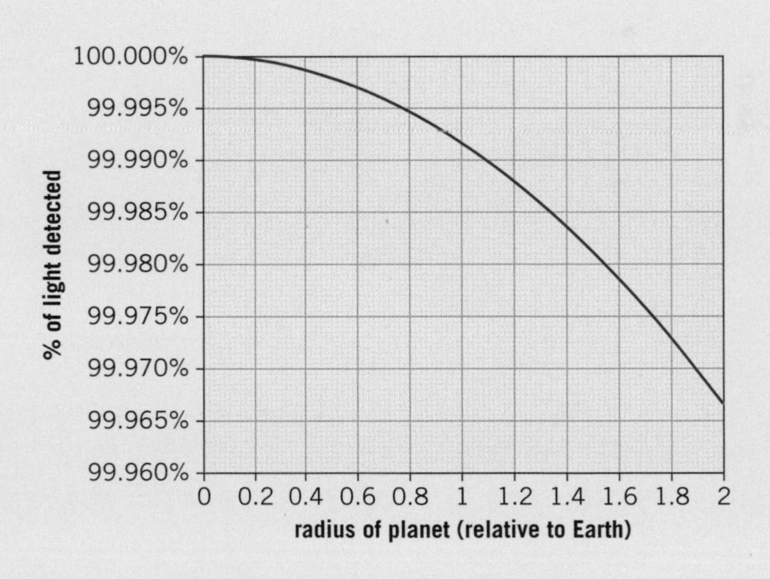

13. Use these graphs to estimate the size of the exoplanet in graph 2 on the previous page. [3]

14. Discuss which planet would be easier to detect. [5]

The search for extra-terrestrial intelligence

The search for extra-terrestrial intelligence (SETI) is trying to detect signals from intelligent life in our galaxy.

15. Suggest some of the problems that might make it difficult to detect communications from an extra-terrestrial civilization. [4]

16. Some scientists are concerned that there are dangers associated with SETI. Discuss the dangers and benefits of trying to communicate with extra-terrestrial life. [6]

▼ These giant radio telescopes search for extra-terrestrial signals

Index

Index headings in **bold** indicate key terms; page numbers in *italics* indicate illustrations/caption text.